Daniel Evans Jones

**Elementary Lessons in Heat, Light and Sound**

Daniel Evans Jones

**Elementary Lessons in Heat, Light and Sound**

ISBN/EAN: 9783743393448

Manufactured in Europe, USA, Canada, Australia, Japa

Cover: Foto ©berggeist007 / pixelio.de

Manufactured and distributed by brebook publishing software (www.brebook.com)

Daniel Evans Jones

**Elementary Lessons in Heat, Light and Sound**

# ELEMENTARY LESSONS

IN

# HEAT, LIGHT, & SOUND

BY

D. E. JONES, B.Sc. (LOND.)

PROFESSOR OF PHYSICS IN THE UNIVERSITY COLLEGE OF WALES
ABERYSTWYTH

ILLUSTRATED BY NUMEROUS ENGRAVINGS

London
MACMILLAN AND CO.
AND NEW YORK
1891

# PREFACE

THESE Lessons are intended to serve the purpose of introducing beginners to the study of Experimental Physics. My object has been, not only to provide a certain amount of useful information, but also to give an idea of the methods by which this information can be most directly obtained. The course has therefore been made experimental, and such brief instructions are given as are required for making or using the apparatus and carrying out the experiments. I have repeatedly tried and modified each experiment so as to present it in a simple form and avoid the more usual causes of failure.

I am well aware that many educational authorities hold that teachers of science (more especially in schools) should confine their instruction to the principles of the subject, without entering into details of manipulation or methods of experiment. To the teacher who aims chiefly at getting 'results' this view readily recommends itself: it saves trouble and expense, and enables him to devote more time to laws and generalisations. Unfortunately the results thus obtained are not of great value. A schoolboy may be taught to repeat glibly certain forms of words respecting the conservation of energy or the atomic theory; but, until he has acquired considerable familiarity with the behaviour and properties of bodies, the words convey no clear idea to his mind. My own experience in teaching has led me to the conclusion that students who come to college with an elementary knowledge of science acquired in this way are very unsatisfactory material to work with. They have been accustomed to get their knowledge at second-hand from their teacher or their book; and they find an experimental course more troublesome, more tedious, and apparently more uncertain. They are easily discouraged and do not see how much can be learned from their own failures. The method fails to bring out one of the chief advantages of science as an educational subject—the training in the habit of observation and of learning from things at first-hand.

In the methods of reasoning, as well as in the choice of words and subject-matter, I have earnestly endeavoured to be as simple and clear as possible. But I have made no attempt to shirk the necessary difficulties that even the beginner must, sooner or later, meet. There still remains much truth in Rousseau's words: 'Parmi tant d'admirables méthodes pour abréger l'étude des sciences, nous aurions grand besoin que quelqu'un nous en donnât une pour les apprendre avec effort.'

The paragraphs in the book which are printed in smaller type may be omitted on the first reading as being of less importance or greater difficulty. Most of the illustrations have been engraved with great care by Mr. J. D. Cooper from my own sketches and photographs of simple apparatus. Others have been borrowed, and for these I am mainly indebted to the courtesy of Messrs. Macmillan. The questions at the ends of the chapters are taken partly from the elementary papers of the Science and Art Department, and partly from my *Examples in Physics.*

My warmest thanks are due to Mr. B. B. Skirrow, B.A., of the Mason Science College, Birmingham, and to Mr. D. S. Macnair, B.Sc., Ph.D., of the People's Palace Technical Schools (E.), for their kind advice and assistance in seeing the book through the press.

D. E. JONES.

UNIVERSITY COLLEGE OF WALES,
ABERYSTWYTH, *March* 1891.

# CONTENTS

## HEAT

| CHAP. | | PAGE |
|---|---|---|
| 1. | TEMPERATURE | 1 |
| 2. | THERMOMETERS | 7 |
| 3. | EXPANSION OF SOLIDS | 14 |
| 4. | EXPANSION OF LIQUIDS | 22 |
| 5. | EXPANSION OF GASES | 27 |
| 6. | SPECIFIC HEAT AND CALORIMETRY | 39 |
| 7. | CHANGE OF STATE—FUSION AND SOLIDIFICATION | 48 |
| 8. | CHANGE OF STATE—VAPORISATION AND CONDENSATION | 56 |
| 9. | HYGROMETRY | 72 |
| 10. | TRANSMISSION OF HEAT—CONDUCTION | 79 |
| 11. | TRANSMISSION OF HEAT—CONVECTION | 89 |
| 12. | TRANSMISSION OF HEAT—RADIATION | 97 |
| | ANSWERS TO EXAMPLES | 111 |

## LIGHT

| | | |
|---|---|---|
| 1. | INTRODUCTION, SHADOWS, ETC. | 115 |
| 2. | PHOTOMETRY | 123 |
| 3. | VELOCITY OF LIGHT | 131 |

## CONTENTS

| CHAP. | | PAGE |
|---|---|---|
| 4. | REFLECTION OF LIGHT—PLANE MIRRORS | 135 |
| 5. | SPHERICAL MIRRORS | 148 |
| 6. | REFRACTION OF LIGHT | 169 |
| 7. | LENSES | 191 |
| 8. | OPTICAL INSTRUMENTS | 212 |
| 9. | DISPERSION AND COLOUR | 217 |
| | ANSWERS TO EXAMPLES | 224 |

## SOUND

| | | |
|---|---|---|
| 1. | INTRODUCTION—VIBRATORY MOTION | 229 |
| 2. | WAVE-MOTION | 236 |
| 3. | TRANSMISSION OF SOUND—ITS VELOCITY | 242 |
| 4. | REFLECTION OF SOUND | 250 |
| 5. | PITCH AND MUSICAL INTERVALS | 255 |
| 6. | TRANSVERSE VIBRATIONS OF STRINGS | 263 |
| 7. | RESONANCE | 269 |
| 8. | VIBRATION OF AIR-COLUMNS—ORGAN-PIPES | 272 |
| | ANSWERS TO EXAMPLES | 280 |

# NOTE ON THE
# METRIC SYSTEM OF WEIGHTS & MEASURES

THE weights and measures used in this book are for the most part those based upon the French or Metric System. In this system the standard of length is the length of a certain bar of platinum preserved in Paris. This is called a *metre*, and is somewhat more than a yard (39.37 inches). It is divided into tenths, hundredths, and thousandths, which are called *decimetres*, *centimetres*, and *millimetres* respectively. Both in measuring and in calculating this decimal method of division is much more convenient than our way of dividing up a yard into 3 feet, each equal to 12 inches and so on. We shall use as our unit of length the *centimetre*. The relation between the centimetre and the inch can be seen from the accompanying figure.

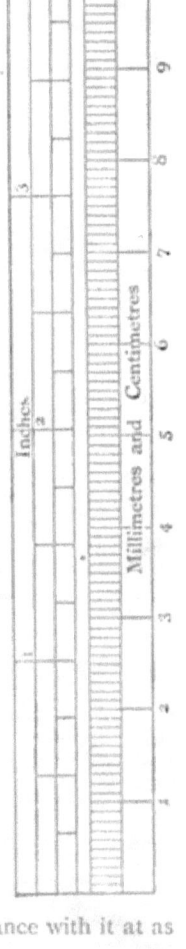

The unit of area is a *square centimetre*, and the unit of volume a *cubic centimetre* (commonly abbreviated into ' 1 c.c.'). The volume of a cubic decimetre is called a *litre* : since 1 decimetre = 10 centimetres, it follows that a litre (or cubic decimetre) contains 1000 c.c. A litre is somewhat more than a pint and three-quarters.

The standard of weight—or rather of mass—is called a *kilogramme*, and is a little more than 2¼ lbs. It is divided into 1000 parts called *grammes*. We shall use the *gramme* as our unit of mass.

The British standard pound bears no simple relation to the units of length and volume used in this country ; but the original standard kilogramme was so constructed as to have a mass equal to that of a cubic decimetre (or litre) of water at a temperature of 4° Centigrade. Thus 1 c.c. (one-thousandth of a cubic decimetre) of water at 4° C. weighs exactly one gramme (one-thousandth of a kilogramme).

A little experience will show the student how convenient it is to have a definite and simple relation between the unit of volume and the unit of mass : he will also find that by using a decimal system we avoid a number of troublesome calculations, such as those required for reducing miles to inches and pounds to ounces. The system is now generally used in scientific books, and by scientific men in all countries : hence it is advisable that the student should acquire some acquaintance with it at as early a stage as possible.

# HEAT

## CHAPTER I

### TEMPERATURE

**1. Hot and Cold Bodies.**—When we stand in front of a fire on a cold day we experience a pleasant sensation of warmth. The same sensation is produced, in a more or less pleasant degree, when we touch objects in the neighbourhood of the fireplace which have been warmed by the fire. We attribute the sensation, in the first case directly, and in the second case indirectly, to the heat given out by the fire, and objects which produce this feeling when we touch them are said to be *hot*. Objects which have not been warmed by the fire will generally produce a sensation of an opposite kind, and we say that they feel *cold*. In ordinary language we frequently use the same word to denote the effect produced and the cause which produces it: thus we speak of the sensation of heat. Whatever heat may be, we may for the present regard it as something that produces in us the sensation of warmth. Let us now consider more closely what we mean by the terms 'hot' and 'cold,' and how we can best find out which of two bodies is hotter than the other.

You will perhaps say that this can easily be found out by touching each of them with the hand and finding out which of the two feels the hotter. Now try this in a room heated artificially, say a hothouse which is heated by steam or hot-water pipes. The pipes themselves, and perhaps the object nearest to them, will feel very hot, and we may naturally assume that they are really hotter than other objects. But these latter produce sensations which differ among themselves in a surprising manner. The wooden shelves on which the flower-pots are arranged do not appear much warmer than the hand itself, whereas the iron brackets on which they are supported feel uncomfortably hot; a piece of slate or iron in a far corner of the room may even appear hotter to the hand than a wooden shelf just above the heating apparatus or a strip of matting wound round the pipes, so that nearness to the source of heat cannot be the cause of the difference in behaviour. Before concluding that the iron and slate are really warmer than the wood and matting we ought to extend our observations and examine them more closely, so as to see whether there is any source of error.

Try a cold room next, say an ordinary dwelling-room on a cold day and without any fire. Touch the various articles in the room as you walk round and note the sensations which they produce in the hand. The table and chairs appear somewhat cold, the mantelpiece (if of slate or marble) colder, and the fireplace and fire-irons colder still. The carpet and table-cloth scarcely produce any sensation of cold at all, and if you bury your hand in a thick soft hearthrug it will appear warm to the touch after a while. The naked soles of the feet, which are ordinarily preserved from extremes of heat and cold by stockings and leather soles, are even more sensitive than the hands. You can readily test this by drawing a hearthrug near to a slate or tiled hearth and standing bare-foot partly on each: the rug scarcely produces any sensation of cold, but the hearthstone is uncomfortably cold, and the floor between slightly cold. And in this case we have no good reason for believing that the stone is really colder than the wood, or the wood than the rug.

**2. Conduction of Heat.**—If you take a poker and push the point of it well into a fire, you will find that after a while the handle begins to get warm; the heat travels along the poker by a process which is called conduction. You can observe the same thing in a teaspoon partly immersed in hot water, and if you take two different teaspoons—a common one made of Britannia metal and a silver one—and put them in the same cup of hot water, you will find that the handle of the silver spoon gets hot more quickly than the other. The silver conducts heat more rapidly than the alloy, or is said to be a better conductor. Metals are generally good conductors of heat, and stone, slate, and marble are fairly good; wood is a bad conductor, and wool, flannel, and fur worse. Now when you take hold of a warm iron rod heat at once begins to flow from the hot body (the iron) to the colder body (your hand) which is in contact with it; and, since iron is a good conductor of heat, there is also a rapid flow of heat from all parts of the rod toward that part which is cooled by contact with the hand. Thus there is a rapid transference of heat from the iron to your hand; your hand gets warm quickly and you say that the iron 'feels hot.' An equally hot wooden rod would not 'feel' nearly as hot to the hand: wood is such a bad conductor that the part of the rod first cooled by contact with the hand would practically remain cool. And so when you touch a cold body; if it is a good conductor, iron or stone, heat flows from your hand and is rapidly conducted away into colder parts of the body. But if it is a bad conductor, such as flannel, most of the heat taken from your hand remains in the part of the flannel which is in contact with the hand, and so it soon begins to feel warm. Thus the heat-sensation which you experience when you touch a body with your hand depends not only upon whether the body is hot or cold, but also to some extent upon its conducting power. It also depends upon the previous state of the hand, as can be shown by the following experiment.

EXPT. 1.—Take three large basins and fill the first with ice or a freezing mixture of ice and salt. Into the second pour tap-water or water slightly warmed, and into the third water as hot as the hand can bear. Plunge your left hand into the ice and your right into the hot water: after holding them there for a few minutes put both hands in the middle basin. The water in it will, by contrast, appear cold to your right hand, while, at the same time, it feels warm to the left hand.

**3. Temperature.**—In scientific language a hot body is said to have a high temperature, while a cold body is said to have a low temperature. It is important that you should understand exactly what this term temperature means. It is not the same thing as heat. Nor is it a quality of any particular body, for a body may be cold at one time and hot at another: it is simply a state or condition of a body. Temperature is sometimes defined as being the 'state of a body with respect to sensible heat;' but we had better from the outset avoid the use of any terms which would seem to indicate that there are different kinds of heat.

When two water cisterns at different levels are connected by a pipe, the difference of level produces a pressure which causes the water to flow from the higher cistern to the lower one. So when a hot body is placed in contact with a colder body heat flows from the hot body to the colder one. Here the difference in temperature corresponds to the difference in level of the water cisterns, and the heat flows from the body at a higher temperature to that at a lower temperature, just as water tends to flow from a higher to a lower level. We are thus led to the following definition:—

*Temperature is a condition of bodies that determines which of two bodies when placed in contact will part with heat to the other.*

The body which parts with the heat is said to have the higher temperature; if there is no transference of heat the two bodies are said to be at the same temperature.

**4. Effects of Heat—Expansion.**—The principal effects produced by applying heat to a body are (1) change of size, (2) change of temperature, and (3) change of state. We shall begin by observing how heat causes solids, liquids, and gases to expand, and then go on to see how this expansion can be used as a means of measuring temperature.

Fig. 1.

EXPT. 2.—Expansion of Solids.—This can be observed by taking a brass ball and a ring (Fig. 1) which fits it loosely, so that the ball can just pass through the ring when both are cold. After the ball has been heated over a spirit-lamp or in the flame of a Bunsen burner it will no longer pass through the ring but will rest upon it, thus showing that it has expanded. As the ball cools it contracts to its original size, and the ring, being warmed by contact, also expands a little: so that after a few minutes the ball drops through.

EXPT. 3.—*Expansion of Liquids.*—Fit a small[1] flask with a cork and long narrow glass tube. Fill the flask with water coloured by a little litmus tincture or solution of indigo; push the cork in so as to force the coloured water half-way up the tube and see that no air is enclosed below the cork. Make a movable index out of a piece of card or stiff paper by cutting two slits near the ends and slipping it over the tube, as in Fig. 2. On immersing the flask in lukewarm water its contents will expand and the column of coloured water in the tube will gradually ascend.

Liquids expand with heat more rapidly than solids, and air and other gases more rapidly than either.

EXPT. 4.—*Expansion of Gases.*—The expansion of air can be observed in a simple way with a flask (empty) fitted as above. Hold the flask upside down with the open end of the tube just dipping under water; the heat of the hand applied to the flask is enough to make the air expand, as can be seen by the bubbles which escape through the water. To fit up the instrument for permanent use take a bottle with a fairly broad bottom, pour a little coloured water into it, and by means of a good thick cork fix the tube in the neck of the bottle with the end dipping under the water (Fig. 3). In the side of the cork cut a slit so as to allow the air to pass

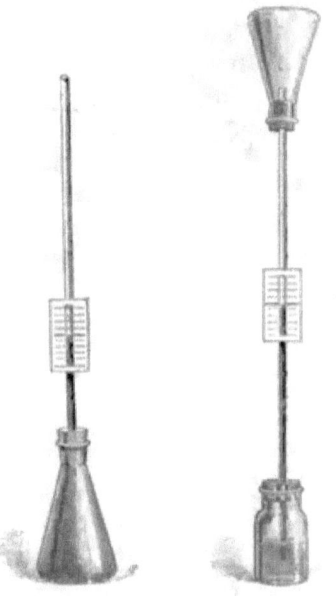

Fig. 2 [1/10].  Fig. 3 [1/10].

[1] In fitting up apparatus for an experiment the exact size will depend largely upon whether it is to be used for private observation or class demonstration. In order to avoid unnecessary details in the text, many of the figures are drawn to scale. The scale is either marked on the engraving or beneath it: thus Fig. 2 is one-tenth [1/10] of the natural size. The sizes thus indicated are generally those suitable for class work.

freely in and out of the lower bottle. Place your hand on the flask, or heat it gently by holding a lighted match near it, until about a dozen bubbles of air have been expelled; the air will contract on cooling, and the column of liquid will ascend about half-way up the tube, which should be provided with a movable index as before. The instrument now forms a fairly sensitive thermoscope; by watching the movements of the column of water you can find out when heat is imparted to, or taken away from the air in the flask. Thus you can tell whether a given sample of water is hotter or colder than the air; for you can pour a few drops of the water upon the inverted bottom of the flask, and if the temperature of the water is higher than that of the air, expansion will occur and the column will descend and *vice versa*. Or again, by moving the thermoscope from one room to another you can compare the temperatures of a series of rooms.

# CHAPTER II

## THERMOMETERS

**5.** What we have called an air-thermoscope (Art. 4) is sometimes called an air-thermometer, but it does not deserve to be dignified with this name, for a **thermometer** (as its name implies) ought not only to *indicate* changes of temperature but also to measure them. Our air-thermoscope does not always give the same indication when exposed to the same temperature; its indication, *i.e.* the height at which the column of water stands, depends upon the pressure of the atmosphere, which varies from time to time. For this and other reasons it would not be easy to use it (in the form described) for measuring temperature. Liquids are not affected by changes of atmospheric pressure as gases are. A rough kind of thermometer can be made by blowing a bulb at one end of a narrow glass tube and filling it with coloured water or some other liquid: the liquid should then be warmed and the open end of the tube sealed. But how is the thermometer to be graduated? You might affix a paper or cardboard scale to it and divide this into a convenient number of equal parts, say twenty or a hundred; but you would find if you made another thermometer in the same way that the two would not give the same indications at the same temperature. Your graduation being an arbitrary one, the indications of one instrument cannot be compared with those of another. A more scientific way would be to begin by choosing two standard temperatures as 'fixed points' to which all thermometers could be referred. The fixed points always adopted are the temperature at which ice melts and that at which water boils. If you immerse a thermometer in melting ice, or in a mixture of ice and water, you will find that it always registers the same temperature; this is called the melting-point of ice, and is taken as the 'lower fixed point' of the thermometer. The temperature at which water boils varies slightly with the atmospheric pressure, but for a given pressure it is constant, so that if we fix upon a certain pressure as the **standard atmospheric pressure**, then the boiling-point of water will give us a second or 'higher fixed point' for our thermometer. These two fixed points are very convenient standards of reference, for both water and ice can always be obtained in a state of purity.

Thermometers are never filled with water: a water-thermometer could only be used through a very limited range of temperature; it would be impossible to mark the fixed points, and even between these the expansion of water is very irregular. Mercury (quicksilver) can be used through a much wider range of temperature; it expands more regularly than other liquids, and for these and other reasons it is almost always preferred as a thermometric fluid.

**6. Construction of a Mercury Thermometer.**—A bulb is blown at one end of a capillary glass tube, *i.e.* a tube with a very fine bore. On account of the contained air and the fineness of the bore the bulb cannot be filled by pouring mercury down, but if you gently warm it so as to expel some of the air,

Fig. 4. THERMOMETERS.

and then dip the open end of the tube under mercury, a little of this will be sucked up as the air cools and contracts. The process of heating and alternate cooling must be repeated several times until the bulb and tube are filled with mercury; finally the whole is carefully heated, so as to drive off any contained air and moisture. While the mercury is still warm the end of the tube is closed by cautiously directing a blow-pipe flame against it, so as to fuse the glass; a vessel closed in this way is said to be **hermetically sealed**. The thermometer should now be allowed to cool very slowly, and should be put aside for several days before the fixed points are marked. After the bulb is blown it does not contract at once to the size which it finally assumes; the greater part of the contraction may occur in the first hour or so, but the bulb goes on shrinking gradually for a long time after. Good thermometer-makers keep a stock of filled tubes, and do not graduate them until about a year or so after they are filled.

7. **Marking the Melting-point.**—In order to determine the lower fixed point the bulb and part of the stem of the thermometer must be covered with melting ice or snow. Put the pounded ice or snow in a large glass funnel supported on a tripod (Fig. 5); push a pencil or penholder into the ice so as to make a hole into which the thermometer is to be introduced, and pack it round with finely-pounded ice. After it has stood for a few minutes, readjust it so that only the top of the mercury column can be seen above the ice; leave it for a quarter of an hour, and then mark off the height of the column by making a fine scratch on the glass with a diamond point or a sharp three-cornered file.

Fig. 5.

8. **The Boiling-point.**—In determining the upper fixed point the bulb and as much as possible of the stem are immersed in a current of steam. The apparatus used is shown in Fig. 6. The steam arising from the boiling water ascends through the inner of the two concentric cylinders, descends (as shown by the arrows) through the annular space between them, and escapes into the air by the spout C. The thermometer is fixed by a cork in a hole in the lid, and should be pushed down until the top of the mercury column is only

just visible above the cork. After standing for a quarter of an hour (or until the mercury has stopped rising), the height is noted. At the same time the barometer should be read: if its height is 76 centimetres (see p. 29) the pressure is that which is taken as the **standard atmospheric pressure**, and the boiling-point is exactly the temperature which is taken as the upper fixed point for thermometers. If the pressure is greater, the boiling-point is higher and *vice versa;* the corrections to be made for these changes will be understood after reading Chapter VIII. It is important to expose as much as possible of the stem, as well as the bulb, to the steam: the bulb should not dip into the water, for the temperature of the boiling water itself is somewhat uncertain.

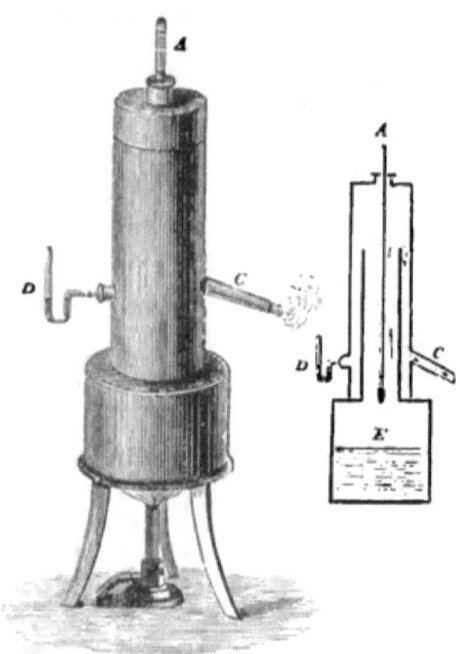

Fig. 6

**9. Graduation—The Centigrade Scale.**—The interval between the two fixed points can now be divided up into a number of parts of equal volume called degrees. In thermometers intended for scientific use the number adopted is one hundred, the lower fixed point being marked 0° and the upper 100°. Thermometers so graduated are called **Centigrade** thermometers, and the **Centigrade Scale** is the one which will be used throughout this book. The graduation is effected by etching it on the glass stem of the instrument: this is covered with a thin coating of wax, upon which the divisions and numbers are marked with a needle point so as to lay bare the glass, after which the marks are **etched** or bitten into the glass with hydrofluoric acid.

The graduation can be extended above the boiling-point (100°) and below the freezing-point (0°). Temperatures below the freezing-point are indicated by a minus sign (−) prefixed: thus '− 10° C.' is read 'minus ten degrees Centigrade,' and indicates a temperature of ten degrees Centigrade below the freezing-point of water.

**10. The Fahrenheit Scale.**—On the Continent (especially in France) the Centigrade scale is used not only for scientific purposes but universally. In England the Fahrenheit Scale is still in popular use. In this scale the interval between the two fixed points is divided into 180 equal parts or degrees. Since the same interval of temperature is divided into 100 degrees in the Centigrade scale, it is clear that one degree Fahrenheit is equal to five-ninths of a degree Centigrade, and a degree Centigrade is equal to nine-fifths of a degree Fahrenheit. Further, the zero (or 0° mark) on the Fahrenheit scale is not placed at the melting-point of ice but 32 degrees Fahrenheit lower; this is about the temperature given by a mixture of ice and salt, and was erroneously supposed by Fahrenheit to be the lowest temperature which could be attained. Thus the melting-point of ice is marked as 32° in the Fahrenheit scale and the boiling-point of water as 212° (32° + 180° = 212°). The relation between the two scales is shown in Fig. 7, and the following examples illustrate the method of conversion from one scale to the other.

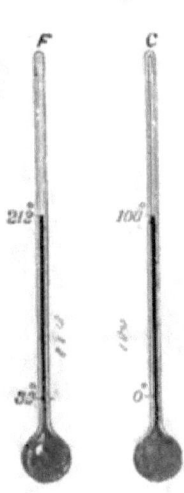

Fig. 7.

**11. Conversion of Scales.**—*Ex.* 1. Find the Centigrade temperature corresponding to 77° F.

> 77° F. is 77 − 32 = 45 Fahrenheit degrees above the freezing-point or Centigrade zero, and this is equivalent to $45 \times \frac{5}{9} = 25$ Centigrade degrees. The required temperature is therefore 25° C.

*Ex.* 2. What is the Fahrenheit temperature corresponding to 15° C.?

> 15 Centigrade degrees are equivalent to $15 \times \frac{9}{5} = 27$ Fahrenheit degrees, and since the Centigrade zero counts as 32° F., the required temperature is 32° + 27° = 59° F.

The rules for conversion of scales may be stated thus :—

To change from Fahrenheit to Centigrade.—Subtract 32, multiply the remainder by 5 and divide by 9.

To change from Centigrade to Fahrenheit.—Multiply by 9, divide by 5 and add 32.

**12. Advantages of the Mercurial Thermometer.**—The convenient form and size of the mercurial thermometer, its cheapness, the ease with which it can be read and the fact that the readings give the temperature directly without any corrections, *e.g.* for barometric height,—these are all points in its favour. By making the capillary bore of the tube very fine and the bulb comparatively large, it can be made almost as delicate as we please ; *e.g.* thermometers can be made to measure the one-hundredth part of a degree Centigrade. The special advantages of mercury as a thermometric fluid are :—

(1) The large range of temperature which it allows. (Mercury solidifies at $-38°·5$ and boils at $357°·9$.)
(2) The regularity of its expansion.
(3) It is opaque ; this enables us to follow the movements of a fine thread of the liquid with ease.
(4) Its specific heat (see Art. 40) is low (0·033) and it is a good conductor of heat ; thus it rapidly takes the temperature of the medium in which it is placed.

**13. Errors and Corrections**—(1) The gradual rise of the zero-point.—This is due to the slow contraction of the bulb and has been referred to in Art. 6. Every thermometer used for accurate work should from time to time have its zero-point redetermined as described in Art. 7. The requisite correction can then be applied throughout the scale ; *e.g.* if the new zero-point is found to be $0°·8$ this amount must be subtracted from all readings, so that a reading of, say, $18°·4$ would indicate a temperature of $17°·6$.

(2) **Temporary lowering of the** zero-point.—This occurs when the instrument has been subjected to a high temperature, as in determining the boiling-point of a liquid ; the bulb does not contract at once to its normal size, so that for some hours afterward the readings of the thermometer are too low.

(3) Error due to exposure of the stem.—This is a frequent source of error in determining high temperatures, *e.g.* in finding the boiling-points of liquids. It is not sufficient to immerse the bulb alone in the vapour of the liquid, for in this case the whole of the mercury is not at the same temperature ; a portion of it (that contained in the exposed part of the stem) is at a lower temperature, approximately that of the surrounding atmosphere, and in consequence the reading will be too low. The most satisfactory way of dealing with this error is to remove it altogether by subjecting the whole of the mercury-column to the temperature to be measured (see Art. 8).

(4) **Errors due to position and pressure.**—A thermometer used in a horizontal position will give higher readings than when it is vertical, on account of the removal of the hydrostatic pressure due to the mercury column: in the vertical position this tends to compress the mercury and dilate the bulb. Thermometers exposed to great pressure give readings which are too high on account of compression of the bulb.

### EXAMPLES ON CHAPTER II

1. On a certain day I found that the temperature of the air rose through 13.5 degrees Fahrenheit between sunrise and mid-day: through how many degrees would a Centigrade thermometer have risen?

2. Find the temperatures on the Centigrade scale corresponding to the following: 59° F., 104° F., 185° F., 203° F.

3. Find the temperatures on the Fahrenheit scale corresponding to the following: 90° C., 80° C., 30° C., −5° C.

4. Express on the Fahrenheit scale the temperatures at which mercury solidifies (−38°.5 C.) and boils (357° C.).

# CHAPTER III

## EXPANSION OF SOLIDS

**14. Linear Expansion of Metals, etc.**—The following is one of the simplest ways of observing the increase in length of a metal bar when it is heated.

EXPT. 5.—Lay a flat bar of iron, about 1 ft. long, across two wooden blocks, as in Fig. 8. Under one end (between it and

Fig. 8 [1/8].

the block) place a fine sewing-needle at right angles to the length of the bar. On the other end place a weight, so as to keep it fixed, and make the bar bear well down upon the needle. Attach a light straw pointer by sealing-wax to the eye end of the needle, and at right-angles to it.

Heat the bar with a spirit-lamp or Bunsen burner: as it expands the free end moves forward and rolls the needle round.

Since the needle is small, a very slight movement of the end will make the pointer move through a sensible arc.

Repeat the experiment with strips of glass (which should be heated cautiously), copper, zinc, etc.

**15. Unequal Expansion of Metals.**—Different metals do not expand equally when heated. The following experiments show that brass expands more than iron when both are heated through the same interval of temperature.

EXPT. 6.—Cut off from a thin iron rod a piece about 6 inches long and file the ends square. Cut a "gauge" out of thick sheet-brass, so that the iron rod just fits into it lengthways when cold (Fig. 9). By filing the inside of the gauge or tapping it with a hammer on the outside you can adjust it so that it grips the iron rod firmly. Suspend the gauge and rod in a sauce-pan or tin dish filled with water, and heat the water to boiling. You will find that the rod soon drops out, thus showing that the brass expands more than the iron.

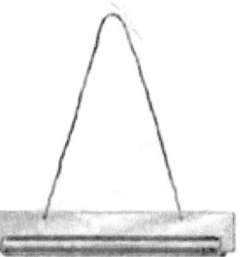

Fig. 9 [1/8].

EXPT. 7.—Take two thin strips of iron and brass respectively: solder or rivet them together, hammer the compound strip

Fig. 10.

straight, and then heat it by holding it over a fire or passing it through a gas-flame. On account of the unequal expansion of the metals the strip bends, the more expansible metal (brass) being on the outside of the curve (Fig. 10).

**16. Coefficients of Expansion.**—Experiment shows that when a metal bar is heated it expands uniformly, and the expansion or increase in length is proportional (1) to the original length of the bar, (2) to the interval of temperature through which it is heated, and (3) to a certain fraction which depends upon the nature of the substance, and which is called the coefficient of linear expansion of that substance.

The coefficient of linear expansion of copper is about $\frac{1}{59000}$ or 0·000017. If a bar of copper 30 inches in length at 0° be heated to 1°, its length will become $30\left(1 + \frac{1}{59000}\right)$ inches or $30(1 + 0·000017)$ inches; and, since the expansion is proportional to the rise of temperature, at 10° its length will become $30\left(1 + \frac{10}{59000}\right)$ inches or $30(1 + 0·000017 \times 10)$ inches, and so on.

The coefficient of linear expansion of a substance may be defined in any of the following ways :—

Definition 1.—If a bar of any substance be heated through 1°, its length will increase by a certain fraction, and this fraction is called the coefficient of linear expansion of the substance.

Definition 2.—The coefficient of linear expansion is the ratio of the increase of length produced by a rise of 1° to the original length.

Definition 3.—The coefficient of linear expansion is numerically equal to the increase in length produced in unit length of the substance by a rise in temperature of 1°.

Let $a$ denote the coefficient of linear expansion of a body whose length at 0° is $l$: starting with any one of the above definitions you will easily see that the expansion produced by heating the body to $t°$ will be $lat$, and that its length $l'$ at $t°$ will be given by the equation

$$l' = l + lat = l(1 + at) \qquad . \qquad . \qquad . \qquad (1)$$

You will find examples of the use of this equation at the end of the chapter, and you should make yourself thoroughly familiar with it.

APPROXIMATE COEFFICIENTS OF LINEAR EXPANSION.

| | |
|---|---|
| Glass | 0·0000086 |
| Platinum | 0·0000086 |
| Iron | 0·000012 |
| Copper | 0·000017 |
| Brass | 0·000019 |
| Zinc | 0·000029 |

**17. Examples and Applications of Expansion.**—You will see by the above table that glass and platinum have the same coefficient of expansion. These two substances expand and contract at the same rate, so that if a platinum wire is fused into a glass tube the joint will remain good on cooling. It is mainly on this account that whenever a metallic wire has to be fused through the side of a glass vessel platinum is always used. If you try to fuse a copper wire into a piece of glass tubing, you will find that on cooling the copper contracts more rapidly than the glass, and so a crack is started which makes the joint leaky.

Boiling water may be poured into thin glass vessels such as beakers and flasks without much danger of cracking them. But glass is such a bad conductor of heat that, if boiling water is poured into a thick-bottomed tumbler, the inside of the tumbler expands before the heat has had time to reach the outside, and in consequence of this irregular expansion the glass cracks.

The best way of loosening a glass stopper when it has stuck fast in the neck of a bottle is to make use of the expansion of glass by heat. Hold the neck of the bottle over a small gas-flame, turning it round so as to heat it gently and uniformly: the neck expands before the heat has time to penetrate farther, and after tapping the stopper with a knife or key it can generally be pulled out.

EXPT. 8.—Stretch an iron or platinum wire, two or three feet in length, horizontally; if you have no suitable clips for holding it, support it on bricks with heavy weights placed on top. Heat it by means of a long burner placed underneath, or, better still, by passing an electrical current from a battery through it. As the wire expands it bends or 'sags' under the influence of its own weight: on cooling it tightens up again. This sagging can be observed in telegraph wires, and is more noticeable in summer than in winter.

Standard yard-measures and metre-scales are constructed so as to be of a given length at a definite temperature: thus the metre is defined as being the distance between the ends of a rod of platinum made by Borda, *the rod being at the temperature of melting ice.* A scale which is correct at $0°$ will not be correct at $15°$, for it will have expanded, and its real length will be greater than its apparent length : to make allowance for this we require to know the coefficient of expansion of the material of the scale, and then we can calculate its real length at any temperature according to equation (1) Art. 16. Such corrections have to be applied to all brass and glass scales affixed to barometers, etc.

**18. Force exerted in Expansion and Contraction.**—Great force is required to stretch a bar of iron appreciably, and when such a bar is made to expand by heat a correspondingly large force is exerted against any obstacle which resists its expansion.

If the bars of a furnace were firmly fixed at both ends, they would, in endeavouring to expand, either loosen the masonry on either side, or else the bars themselves would bend. Instead of this, they are fixed at one end only and are free to expand at the other end; or, as in boiler-furnaces, the fire-bars are not fixed at all at the ends, but simply rest upon cross-bars.

Wheelwrights employ the force of contraction in making cart-wheels. When the wooden framework of the wheel is ready it is placed in position on the ground: the tire or iron rim, which is made somewhat smaller in diameter, is heated until it has expanded sufficiently to allow the wheelwright to slip it on. Cold water is now poured over the iron, and as it shrinks it binds the parts of the wheel firmly together.

In making railways, especially if the rails are laid in cold weather, a space of about a quarter of an inch is left between each rail and the next in order to allow for expansion in summer. In all large metal structures engineers endeavour to obtain freedom of movement under changes of temperature. In the Britannia tubular bridge over the Menai Straits this is secured by resting the metal tubes on rollers at one end; and in the huge Forth bridge, which has recently been constructed, the rails on the 1700-feet span are free to slide to the extent of 18 inches.

**19. Compensation of Clocks and Watches.**—The rate at which a clock goes is controlled by the movements of its pendulum, and the time taken by the pendulum to swing backwards and forwards depends upon its length. You can test this with a pendulum consisting of a bullet hung from a string. If the string is a metre (or about 39 inches) in length, the bullet takes just a second to swing from side to side: if you reduce the length of the string to one half, it will swing four times as quickly. Now the metal rod which supports the bob of a pendulum expands and contracts as the temperature rises and falls, so that unless the pendulum is compensated for changes of temperature the clock will go faster in summer than in winter.

If you refer to the table in Art. 16, you will see that the coefficient of expansion of brass is about $1\frac{1}{2}$ times that of iron.

Now let AC (Fig. 11) represent an iron rod and BC a brass rod, the two being connected by a cross bar at C: if the iron rod is made just $1\frac{1}{2}$ times as long as the other, both will expand equally when the temperature rises. Further, if the point A is fixed, the iron rod AC will expand downwards, while the brass rod CB will expand upwards *through the same distance*, so that if A were the point of suspension of a pendulum and B its bob, the length of the pendulum would not be affected by a change of temperature. A pendulum made exactly as in Fig. 11 would be long and awkward; the form commonly used is the 'gridiron pendulum' shown in Fig. 12, in which the dark lines represent iron rods and the lighter lines brass rods.

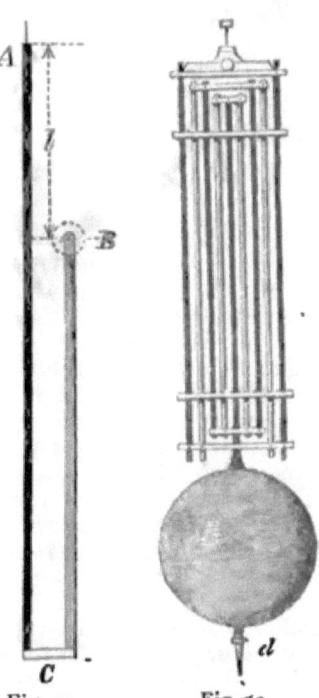

Fig. 11.   Fig. 12.

If you examine a watch, you will find in it a wheel with a heavy rim, which is called a balance-wheel: this swings backwards and forwards and regulates the rate of the watch as a pendulum does that of a clock. In hot weather the wheel expands and the watch tends to go more slowly: in order to prevent this, good watches and chronometers are fitted with compensated balance-wheels, in which the rim is divided into two or three separate pieces (Fig. 13), each of which is fixed to a spoke at one end and is loaded at the free end. Each portion of the rim is made of two metals of unequal expansibility, such as steel and brass, the more expansible metal being outside: this is shown in Fig. 13, where the black segments represent steel and the shaded ones brass. You have learned from Expt. 7 (Art. 15) that such strips curl *inwards* when heated, and in so doing they throw their loaded ends nearer the centre of the wheel. At the same time the direct expansion of the spoke pushes the rim *outwards*. These two effects can be made to compensate one another so as to keep the time of oscillation the same in summer as in winter.

Fig. 13.

**20. Superficial and Cubical Expansion.**—With the necessary change of terms, the coefficients of superficial and cubical expansion may be defined in the same way as the coefficient of linear expansion (Art. 16).

Let ABCD represent a square, each of whose sides is of unit length at 0°. (In the figure each side is an inch long and the area ABCD is a square inch.) Let the coefficient of linear expansion of the material of which the square is made be denoted by $a$: on heating it from 0° to 1° the square will expand so that each side increases in length from 1 to $1+a$. Let A$bcd$ be the expanded square: each of the short lengths B$b$, D$d$ represents the linear expansion ($a$), and the shaded border at the bottom and right hand represents the superficial expansion or increase of area. Consider the strip C$b$: its length is unity, and its breadth is $a$: thus its area is $a$ (or that fraction of an inch), and so is the area of C$d$.

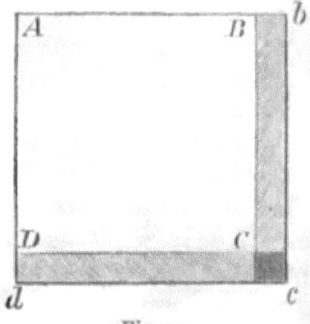

Fig. 14.

*If we neglect the corner-piece* C$c$, its size being very small compared with AC or C$b$, we may say that the increase of area is $2a$, and the increase is produced by raising the temperature of unit area through 1°. Hence (see Def. 3, Art. 16) the coefficient of superficial expansion is $2a$ or is double the coefficient of linear expansion.

This may also be shown in another way. The length of each side of the expanded square at 1° is $1+a$. The area of the square is therefore $(1+a)^2 = 1 + 2a + a^2$. Now $a$—the coefficient of linear expansion—is always a small fraction: the largest coefficient given in Art. 16 is less than 0·00003, so that the largest value of $a^2$ is less than 0·0000000009, and in comparison with the other quantities this is so small that we may neglect it.

Similarly it can be proved that the coefficient of cubical expansion is approximately three times the coefficient of linear expansion. Consider a cube each of whose sides is of unit length at 0° and of length $1+a$ at 1°. The volume of the cube at the latter temperature is

$$(1+a)^3 = 1 + 3a + 3a^2 + a^3.$$

We have already seen that $a^2$ is very small and $a^3$ is smaller still; we may therefore neglect the last two terms. The volume of our cube which was unity at 0° becomes $1+3a$ at 1°, and therefore the coefficient of cubical expansion is $3a$.

### Examples on Chapter III

(The coefficients of expansion used are given on page 16)

1. Find the length at 200° of a zinc rod whose length at 0° is 128 cm.
If we denote by $l_{200}$ the length at 200°, then by equation (1), page 16,
$$l_{200} = l_0(1 + 200a),$$
$$= 128(1 + 200 \times 0.000029),$$
$$= 128 \times 1.0058 = 128.7424 \text{ cm.}$$

2. A piece of brass wire is exactly 3 metres long at 250°: what will be its length at 0°?
Using the same system of notation, we have
$$l_0 = \frac{l_{250}}{1 + 250a} = \frac{300}{1 + (250 \times 0.000019)},$$
$$= \frac{300}{1.00475} = 298.582 \text{ cm.}$$

3. An iron steam-pipe is 40 feet long at 0°: what will be its length when steam at 100° passes through it?

4. What is meant by saying that the coefficient of expansion of steel is 0.000012? Assuming the highest summer temperature to be 40° C., and the lowest winter temperature −20° C., what allowance should be made for expansion in one of the 1700-feet spans of the Forth Bridge?

5. The length of the iron railway bridge across the Menai Straits is about 461 metres. Find the total expansion of this iron tube between −5° and 35°.

6. A copper rod, the length of which at 0° is 2 metres, is heated to 200°: what will its length now be? At what temperature will its length be 200.51 centimetres?

7. An iron yard-measure is correct at the temperature of melting ice: express, as a fraction of an inch, its error at the temperature of boiling water.

# CHAPTER IV

## EXPANSION OF LIQUIDS

**21. Real and Apparent Expansion.**—In the experiment with a flask and tube described in Art. 4 it is not the real expansion of the liquid that is observed. The flask expands at the same time and its capacity increases, so that the observed or apparent expansion of the liquid is less than its real expansion. That the flask does expand can be shown in a very striking way by heating it suddenly. Instead of immersing it in lukewarm water, so as to heat it gradually, plunge it at once into hot water and watch the top of the column in the tube. This will fall as if there were a sudden contraction of the liquid: the contraction soon ceases and is followed by a steady rise of the column. The apparent contraction is accounted for by the fact that the flask expands before the heat has time to pass through the glass and warm the water inside. Thermometers behave in the same way. It is clear that the liquid expands more than the glass, for otherwise there would be no rise of the column.

We have thus to distinguish between the real expansion of a liquid and its apparent expansion as observed in a glass vessel. It can be proved that the real expansion is equal to the apparent expansion together with the cubical expansion of the containing vessel.

**22. Unequal Expansion of Liquids.**—EXPT. 9.—Fit up three glass flasks with tubes as in Expt. 3. The flasks should have a capacity of about 100 c.c., but in any case they should be of the same size and be fitted with tubes of the same bore. Fill one flask with mercury, the second with coloured water, and the third with some other liquid, such as alcohol or common methylated spirit coloured with turmeric. Put in the corks so that the liquids stand at the same height in the three tubes and fix a cardboard scale behind to indicate this height. Place the three flasks side by side in a tin trough or any other convenient vessel and pour into this sufficient lukewarm water to cover the flasks up to the corks. Wait until the expansion has ceased and there is no further rise: when this is the case the flasks and their contents will be all three at the same temperature, and to make sure of this you may stir the water about so as to mix it.

You will now find that all three columns have risen, but to different heights: the mercury has risen least, the water next, and the alcohol or spirit much more than either. At the beginning of the experiment the three liquids were at the same temperature and occupied the same volume: they have been heated through the same interval of temperature and now they occupy different volumes. What you observe is, of course, the apparent expansion, but as the flasks are of the same size and made of the same material, it follows that the observed expansion is less than the real expansion by the same amount in all three cases. We can, therefore, conclude with certainty from this experiment that different liquids expand unequally when heated through the same interval of temperature: in other words, different liquids have different coefficients of expansion.

**23. Measurement of Apparent Expansion.**—If a glass vessel filled with any liquid is warmed, a portion of the liquid overflows; and if you weigh the amount of the liquid expelled and also that which is left in the vessel, you can calculate the coefficient of apparent expansion of the liquid in glass. The vessel used is a spherical or cylindrical glass bulb with a narrow neck, and is called a **weight-thermometer**.

Fig. 15.

EXPT. 10.—To find the coefficient of apparent expansion of mercury. Make a weight-thermometer by blowing a bulb at the end of a piece of thermometer tubing, or seal off one end of a wide tube and draw out the other end into a fine neck. Bend the neck twice at right angles, as shown in Fig. 15. Weigh the empty thermometer. Fill it in the same way as an ordinary thermometer (Art. 6) by dipping the open end or beak under mercury and alternately heating and cooling the bulb until all air is expelled and the mercury fills the bulb and tube completely. Allow it to cool slowly (with the beak still dipping under mercury) to the temperature of the air and note what that temperature is. Clean and weigh a small porcelain crucible and place it under the beak of the thermometer. Warm the bulb and its contents to a convenient known temperature (say 60°) by immersing it in a large beaker of hot water. Weigh the mercury which overflows into the crucible: also dry and weigh the thermometer.

You can improve the above method by cooling the mercury in melting ice to 0° (instead of to the temperature of the air) and heating it to 100° in the boiling-point apparatus (Fig. 6). The following are the results of such an experiment :—

Weight of mercury expelled at 100° . . . . 8.49 gm.
Weight of mercury contained in thermometer at 100° . 545 ,,

The amount expelled (8.49 gm.) represents the increase in volume produced by heating 545 gm. through 100°: one-hundredth of this, or 0.0849 gm., represents the increase of volume for a rise of 1°.

The ratio of this to the original volume (see Def. 2, Art. 16) is the coefficient of apparent expansion, which is therefore equal to 0.0849/545, or 0.0001558.

From this the real expansion of mercury can be found when the cubical expansion of the glass is known. Assuming, according to Art. 20, that the coefficient of cubical expansion of the glass is $3 \times 0.0000086$ or 0.0000258, we find that the coefficient of real expansion of mercury is $0.0001558 + 0.0000258 = 0.0001816$.

**24. Peculiar behaviour of Water.**—When water is gradually cooled it does not contract regularly down to the freezing-point, but begins to expand again before it reaches this temperature.

EXPT. 11.—Make a coil of lead tubing by bending it round a bottle. Close the lower end with a cork; leave the other end sticking up and fit it with a cork and narrow tube. Fill the coil and tube with water: this is best done by sucking up the water before the bottom cork is pushed in. Now cover the coil with ice and watch the water in the tube. At first it falls rapidly; gradually the contraction becomes slower, and after a while the column comes to rest: then it begins to rise, showing that the contraction has been followed by an expansion.

Remove the ice and let the water gradually rise to the temperature of the air: at first it contracts, then stops, and finally begins to expand.

The temperature at which the water occupies the smallest volume is 4°: both below and above this temperature its volume increases. This is expressed by saying that water has its maximum density at 4°.

The density of a substance is defined to be the mass of

unit-volume of that substance. If M be the mass of a body and V its volume, its density is given by the equation $D = M/V$. If we measure mass in grammes and volume in cubic centimetres, the density of a substance will be measured by the number of grammes in a cubic centimetre of that substance. Thus when we say that the density of mercury is 13.6, we mean that 1 c.c. of mercury weighs 13.6 grammes. When a substance expands by heat its density becomes less in proportion as its volume increases. The following experiment shows that hot water is lighter or less dense than cold water.

EXPT. 12.—Boil some water in a flask or saucepan, adding a few drops of ink or indigo solution to colour it. Half fill a beaker with cold water and on its surface float a thin piece of wood. Pour the hot water gently upon this: the wood rises with it but serves to break its fall and prevent it from rushing at once to the bottom of the beaker. The hot water does not mix immediately with the cold water, but being lighter floats above it, and is easily distinguished by its blue colour.

**25. Hope's Experiment.**—Hope showed by the following experiment that water has its maximum density at 4°.

A freezing mixture is applied round the middle of a cylindrical jar (Fig. 16) filled with water at the temperature of the air. Thermometers are inserted through holes in the sides of the vessel, so as to give the temperature of the water in it at the top and bottom. The first effect produced by the freezing mixture is that the lower thermometer falls, whereas the upper one is scarcely affected: this shows that the water sinks to the bottom as it is cooled by the freezing mixture. But the temperature indicated by the lower thermometer does not fall steadily to the freezing-point—*it stops at* 4°. Now the upper thermometer begins to fall, and it *falls right down to* 0°. This shows that water at 0° is lighter than water at 4°, and swims upon its surface.

Fig. 16.

This fact is of great importance in nature, for the process

which we have described is just what takes place in ponds and lakes in winter. As the water cools it becomes heavier and sinks to the bottom, its place being taken by warmer and lighter water. If water contracted regularly down to 0°, this process would go on until all the water in the lake was reduced to 0°, and it would then begin to freeze outright. But the circulation stops as soon as the temperature reaches 4°; after this the colder water floats on the surface and gradually freezes. Ice is lighter than water and is also a bad conductor of heat; thus the sheet of ice formed on the surface of a lake to some extent protects it from further cooling. The deep water of lakes in our climate seldom falls below 4°, so that fish can exist in them through our coldest winters.

### Examples on Chapter IV

1. Distinguish between the real and apparent expansion of a liquid contained in a glass vessel. If the coefficient of apparent expansion of mercury contained in a glass vessel is $\frac{1}{6500}$, while its coefficient of real expansion is $\frac{1}{5500}$, what is the coefficient of cubical expansion of the glass?

2. The volume of a gramme of water being 1 c.c. at 4°, and 1·0169 c.c. at 60°, what is the mean coefficient of expansion of water between these temperatures?

3. Calculate the coefficient of apparent expansion of mercury in glass from the following results of an experiment with a weight-thermometer:—

Weight of mercury expelled at 100° . . . . 10·2877 gm.
Weight of mercury contained in thermometer at 100° . 659 gm.

4. A weight-thermometer which contains a kilogramme (1000 gm.) of mercury at 0° is placed in an oil-bath, and the mercury expelled is found to weigh 20 grammes. Find the temperature of the bath, the coefficient of apparent expansion of mercury in glass being 0·000155.

5. A flask containing ice-cold water is warmed gradually by letting it stand in a room at 10°. How will the volume of the water alter during the process? What would happen if the flask were plunged into hot water so as to heat it more quickly?

6. A pond of water has been cooled just to the point of freezing: would you expect the temperature and density of the water to be the same at the top and bottom of the pond?

7. The coefficient of linear expansion of glass is 0·000008, and a certain glass flask holds exactly 100 c.c. of water at 0°. What will be the volume of the water contained when the flask and its contents are heated to 100°?

# CHAPTER V

## EXPANSION OF GASES

**26. Behaviour of Gases as compared with Liquids and Solids.**—Every boy who has used a pop-gun knows that air can be compressed or diminished in volume by increasing the pressure upon it. In the pop-gun a certain quantity of air is shut off between two corks in a tube : by pressing a rod against one of the corks it is pushed nearer the other ; the volume of the enclosed air is gradually diminished and, at the same time, its pressure gradually increases. At last the pressure becomes great enough to blow out the front cork, and the enclosed air suddenly expands and returns to its original state.

If you fill the pop-gun with water, you know that the water accommodates itself to the form of the tube, as indeed it does to the form of any vessel into which it is poured. But if you cork up the water and try to compress it in the same way as air, you will find that you cannot do it ; either the front cork is pushed out at once, or else the water leaks out through the corks. It is true that water can be slightly compressed by subjecting it to great pressures, but you would require a much stronger instrument than a pop-gun to do this, and a much more delicate means of observing the effect.

Not only are gases more compressible than liquids, but they also show a tendency to expand which liquids do not possess. If you pour a cupful of water into a glass flask, the water lies at the bottom of the flask, and a definite surface separates the water from the air above it. It easily accommodates itself to the shape of the flask, but it shows no tendency to accommodate itself to the size of the flask, *i.e.* to fill it. On the other hand, if you pump out all the air from a flask and then introduce into it a small quantity of air or any other gas, however small the quantity of the gas may be, it will expand and fill the flask entirely. And if the size of the flask could by any means be increased, the gas would still expand and continually fill it.

Thus you see what are the peculiar properties of solids, liquids, and gases. A solid has a definite form and shape, which cannot be altered except by applying force to it. A liquid has no definite form, but takes the form of the vessel which contains it. A gas not only has no definite form, but its bulk or volume can also be readily changed: it can be easily compressed and it also tends to expand so as to fill completely any vessel in which it is placed.

27. **Pressure of the Atmosphere.**—EXPT. 13. Choose a glass tumbler having a smooth rim and fill it to the brim with water. Slide a piece of stiff paper over the rim, so as to cover it and exclude all air-bubbles. Place your hand on top and turn the whole upside down. On removing your hand you will find that nothing happens—the water does not run out.

What is it that supports the water? It is simply the pressure of the atmosphere. This pressure is exerted by the air in all directions—downwards, upwards, and sideways: in the above experiment the air underneath the paper presses it upwards and so prevents the water from falling out. It can be shown by experiment and calculation that the pressure exerted by the atmosphere upon objects at the surface of the earth is about equal to a weight of 15 lbs. per square inch, or somewhat more than a weight of 1000 grammes on every square centimetre. The pressure on the mouth of a pint tumbler would be over 100 lbs. weight—far greater than the weight of the water contained in it. You will have some idea of what this atmospheric pressure amounts to when it is stated that the pressure exerted by the air on a house is generally greater than the weight of the house itself. We will now see how this pressure can be measured.

28. **The Barometer.**—EXPT. 14. Close one end of a clean glass tube about a yard in length, and fill it with mercury. Place your thumb over the open end (taking care that no air is enclosed) and invert the tube in a small basin or glass mortar

containing mercury, keeping your thumb tightly pressed against the open end of the tube until it is well under the mercury. Now remove your thumb. Only a portion of the mercury runs out of the tube; a column of mercury about 30 inches or 76 centimetres in height is supported in the tube by the pressure of the atmosphere. Repeat the experiment with tubes of different lengths and diameters, and measure the height of the column each time; you will find that it does not depend upon the size of the tube provided that the tube is always long enough.

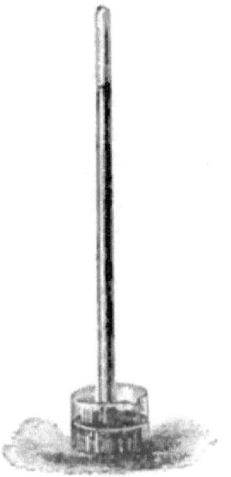

Fig. 17.

The instrument is called a barometer, because it enables us to measure the weight or pressure of the atmosphere. The space in the tube above the mercury is as near an approach to a vacuum or empty space as we can obtain: it is called the Torricellian vacuum, after Torricelli, by whom the experiment was devised. The barometric height varies from time to time and from place to place according to the changes of the atmospheric pressure. We shall adopt a pressure equal to that of a column of mercury 76 centimetres high as the standard atmospheric pressure.

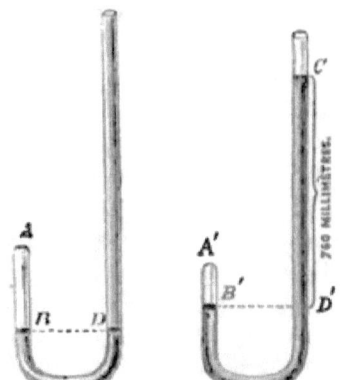

Fig. 18.

**29. Boyle's Law.**—EXPT. 15. Choose a long glass tube of uniform bore; clean it carefully, close one end and bend it as in Fig. 18. The open end should be more than 76 centimetres long. Fasten the tube to an upright board and pour into it a little mercury (less than is shown in the figure), so as to close the bend. By pouring more mercury into the open limb you can find out the effect of varying the pressure upon the air shut off in the closed limb.

Suppose that the height of the barometer is 76 centimetres. When the mercury is at the same level in both limbs of our tube the pressure of the enclosed air is the same as that of the air outside: it is equal to that of a column of mercury 76 centimetres high. Now pour mercury into the tube until the difference of level (D'C) between the closed and open limbs is 76 centimetres; this produces an increase of pressure equal to the original pressure of the enclosed air. You have increased the pressure upon it from one atmosphere to two atmospheres, and you will find that the volume of the air is in consequence reduced to one-half. If you were to increase the pressure to three atmospheres, you would see that the volume could be reduced to one-third, and so on.

By such experiments Boyle was led to discover the following law:—

**Boyle's Law.**—*The volume of a given mass of gas is inversely proportional to its pressure.*

In what is stated above it is supposed that the temperature of the gas is kept constant; we now proceed to consider the effect of heating a gas.

30. **Equal Expansion of Gases**.—We have already seen that heat causes both liquids and gases to expand, and we found from Expt. 9 that different liquids have different coefficients of expansion: the following experiment will show us that this is not true of gases.

EXPT. 16.—Choose two flasks of equal capacity (as in Expt. 9), fitted with corks and narrow tubes. Bend the tubes so that they may dip under water contained in a dish. Fill two equal test-tubes with water and invert them over the open ends of the tubes. Have ready some hot water and a tin trough deep enough to contain the flasks. Before putting the flasks in position fill one of them with coal-gas by upward displacement; or you may fit it with another tube and clip and drive a current of coal-gas through it. Pour hot water into the trough until the flasks are covered up to the corks. The contents of both flasks expand rapidly—far more rapidly than the liquids in Expt. 9. The increase of volume can be measured by the amounts of gas collected in the two test-tubes. You will find that these amounts are equal. And if you repeat the experiment with hydrogen, oxygen, or nitrogen, you will always get the same result.

We learn from this (1) that gases expand more rapidly than liquids; (2) that equal volumes of these gases expand equally when heated through the same interval of temperature—in other words, that they have the same coefficient of expansion.

**31. Charles's Law.**—Experiments made by Charles, and soon afterwards by Gay-Lussac, have established the following law:—

**Charles's Law.**—*The volume of a given mass of gas, kept at a constant pressure, increases by a definite fraction of its amount at $0°$ for each degree rise in temperature.*

This fraction is called the coefficient of expansion of the gas. For all the gases referred to above (air, oxygen, hydrogen, nitrogen) its value is 0·00366, or about $\frac{1}{273}$.

Thus if we take a quantity of gas whose volume at $0°$ is 1 (unity), its volume will become

$$1 + \frac{1}{273} \text{ at } 1°,$$

$$1 + \frac{2}{273} \text{ at } 2°,$$

$$1 + \frac{3}{273} \text{ at } 3°,$$

and

$$1 + \frac{t}{273} \text{ at } t°.$$

Or if we denote by $V_0$ the volume at $0°$ and by $V_t$ the volume at $t°$, then

$$V_t = V_0 \left(1 + \frac{t}{273}\right) \qquad (1)$$

*Ex.* 1.—A certain quantity of oxygen gas occupies a volume of 300 c.c. at $0°$ find its volume at $91°$.

The required volume is given by the equation

$$V_{91} = V_0 \left(1 + \frac{91}{273}\right)$$

or

$$V_{91} = 300 \left(1 + \frac{1}{3}\right) = 300 \times \frac{4}{3} = 400 \text{ c.c.}$$

If we adopt the symbol $a$ to denote the coefficient of expansion, we may put equation (1) into the algebraical form

$$V_t = V_0(1 + at) \quad . \qquad (2)$$

from which we obtain

$$V_0 = \frac{V_t}{1 + at} \qquad . \qquad (3)$$

an equation by which we can find the volume at $0°$ when the volume at $t°$ is given.

With reference to this important law observe the following points :—

(*a*) 'Kept at a constant pressure.'—These words are inserted because the law only holds good if the pressure remains constant : if it varies, the volume also varies according to Boyle's Law (Art. 29).

(*b*) 'For each degree.'—The increase of volume is proportional to the rise of temperature.

(*c*) The coefficient of expansion (which is *defined* by the statements at the beginning of this article) is the *same* for all the gases named. This is in accordance with Expt. 17.

(*d*) 'Of its amount at $0°$.'—Just as a certain pressure (76 cm. of mercury) is adopted as the standard atmospheric pressure, so this temperature ($0°$) is adopted as the standard or normal temperature for measuring the volume of a gas, and the value of the coefficient of expansion is stated with reference to this. If you are given the volume at $20°$ you are not at liberty to say that at $21°$ the volume will have increased by the *same* fraction $\left(\frac{1}{273}\right)$. You must either find by equation (3) its volume at $0°$, and then by equation (2) its volume at $21°$, or you may proceed as follows :—

*Given the volume $V_t$ of a mass of gas at $t°$, to find its volume $V_{t'}$ at $t'°$.*

As in equation (2) we can refer the volumes at both temperatures to the volume $V_0$ at $0°$, thus

$$V_t = V_0(1 + at),$$
and
$$V_{t'} = V_0(1 + at').$$

It follows that

$$\frac{V_{t'}}{V_t} = \frac{1 + at'}{1 + at} \qquad . \qquad . \qquad . \qquad . \qquad (4)$$

$$V_{t'} = V_t(1 + at')/(1 + at).$$

Since $a = \frac{1}{273}$ we may write equation (4) in the form

$$\frac{V_{t'}}{V_t} = \frac{1 + t'/273}{1 + t/273} = \frac{273 + t'}{273 + t} \qquad . \qquad . \qquad (5)$$

which is convenient for calculation (see next article).

**32. Absolute Temperature.**—Charles's law holds good not only for expansion on heating but also for contraction on cooling. Thus if the volume of a certain mass of gas at 0° be 1 (unity) its volume on cooling would become

$$1 - \frac{1}{273} \text{ at } -1°,$$

$$1 - \frac{2}{273} \text{ at } -2°,$$

and
$$1 - \frac{t}{273} \text{ at } -t°.$$

Now suppose the gas to be cooled down to $-273°$: if this were possible, and if the law held good down to this temperature, the volume of the gas would be reduced to zero (for $1 - \frac{273}{273} = 1 - 1 = 0$). This temperature ($-273°$) is commonly called the absolute zero of temperature. A scale of temperatures in which this is taken as the zero is called the absolute scale of temperature, and temperatures reckoned according to this scale are called absolute temperatures. Clearly the absolute temperature corresponding to any given temperature on the Centigrade scale will be found by adding 273 to it. Thus the following are corresponding temperatures:—

| Centigrade Scale. | Absolute Scale. |
|---|---|
| 0° | 273° |
| 1° | 274° |
| 100° | 373° |
| 273° | 546° |
| $t°$ | $T° = 273° + t°$ |

It will now be seen that the numerator and denominator on the right-hand side of equation (5) represent respectively the absolute temperatures corresponding to $t°$ and $t°$ on the Centigrade scale; denoting these by $T'$ and $T$ we may write the equation in the form

$$\frac{V_{t'}}{V_t} = \frac{T'}{T} \qquad \qquad (6)$$

or, *The volume of a given mass of gas is proportional to its absolute temperature.*

*Ex.* 2.—15 litres of air are cooled from 27° to 7°: by how much will the volume diminish?

The absolute temperature corresponding to 27° C. is $273° + 27° = 300°$; and the absolute temperature corresponding to 7° C. is $273° + 7° = 280°$. Let V be the volume of the air at the lower temperature; since the volumes are proportional to the absolute temperatures,

$$\therefore \frac{V}{15} = \frac{280}{300} = \frac{14}{15},$$

or
$$V = 14 \text{ litres.}$$

Thus the volume of the air becomes 1 litre less.

**33. Change of Pressure at** Constant Volume.—We have hitherto supposed the gas to be kept at constant pressure under such conditions that it is free to expand as its temperature rises. Suppose now that instead of keeping the pressure constant we keep the volume constant; what will happen when the gas is heated? It will exert an increasing pressure upon the walls of the vessel in which it is contained, and experiment shows that the pressure increases in the same way and at the same rate as the increase of volume discussed in Art. 31. We have thus to distinguish between two sets of conditions under which a gas may be heated, with the corresponding effects—

| CONDITION. | EFFECT. |
|---|---|
| (1) Constant Pressure. | Increase of Volume (according to Charles's law). |
| (2) Constant Volume. | Increase of Pressure (according to the same law). |

For the second case we may therefore put Charles's law in the following form.

*The pressure of a given mass of gas, kept at constant volume, increases by a definite fraction of its amount at 0° for each degree rise in temperature.*

This fraction is called the coefficient of increase of pressure at constant volume; its value is the same $\left(\frac{1}{273}\right)$ as that of the coefficient of increase of volume at constant pressure. All the equations given above (1-6) apply equally to change of pressure; *e.g.* equation (2) may be written

$$P_t = P_0 (1 + \alpha t). \quad . \quad . \quad . \quad . \quad (7)$$

where $P_0$ denotes the pressure at 0°, and $P_t$ the pressure at $t°$.

*Ex.* 3.—A steel cylinder placed in melting ice is filled with compressed oxygen at a pressure of 42 atmospheres; if the cylinder is now taken out of the freezing mixture and allowed to stand in a room at 26°, what will the pressure become?

Since the pressure increases by $\frac{1}{273}$ of its original value at 0° for every degree rise in temperature, the pressure at 26° will be

$$P_{26} = 42 \left(1 + \frac{26}{273}\right) = 42 \left(1 + \frac{2}{21}\right)$$

$$= \frac{42 \times 23}{21} = 46 \text{ atmospheres.}$$

**34. Air-Thermometers.**—We may here refer to an instrument called the **differential** air-thermometer, a convenient form of which is shown in Fig. 19. The U-tube in the centre contains enough coloured water to fill each limb to a height of a couple of inches. Each limb is connected by a short piece of rubber tubing with a glass tube having a bulb at the end. If one of the bulbs—say the left-hand one—be warmed, the air in it will expand and force the liquid column or index down in the left-hand limb in the U-tube and up in the right-hand limb. The differential air-thermometer gives us an easy and

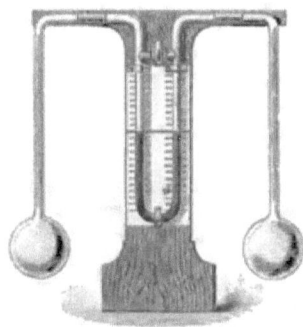

Fig. 19 [1/10].

delicate means of finding whether there is any difference of temperature between two liquids and will be used for this purpose in Expt. 18. If there is any difference, the index moves downwards on the side nearer the warmer liquid; if both liquids are at the same temperature, the index remains at rest.

The cross-tube and stop-cock are not absolutely necessary, but are useful for equalising the pressure in the two bulbs and levelling the index at any time. A simple but serviceable form of differential thermometer can be made from a couple of small glass flasks and a piece of tubing bent six times at right angles, thus— ⌐⌐⌐⌐

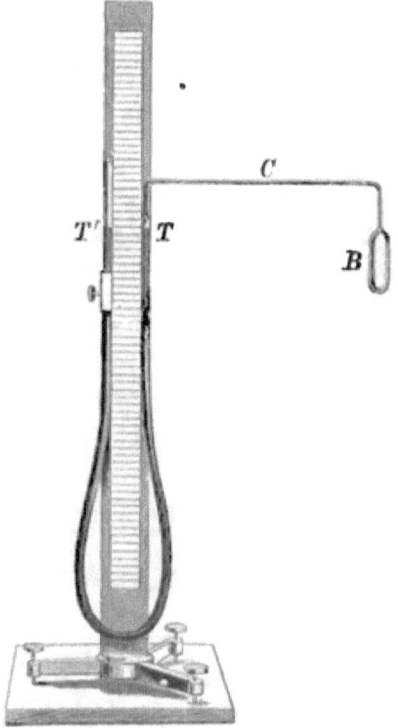

Fig. 20 [1/20].

After what we have said about changes of pressure and volume, you will be able to understand why we cannot use the air-thermoscope (described in Art. 4-5) as a thermometer. The atmospheric pressure changes from time to time, causing alterations in the volume of the contained air independently of changes of temperature; thus when the external pressure increases the column of liquid is forced up the tube just as it is when there is a fall of temperature. And beside all this, the very movement of the liquid column alters the pressure so that it never remains constant.

What is properly called an air-thermometer is shown in Fig. 20. The air is contained in the bulb B, which is connected by a capillary tube C with a wider glass tube T fixed to a vertical stand: T' is a similar tube, which can be moved up and down and adjusted at any height. T and T' are connected below by a strong india-rubber tube, and the U-tube thus formed is filled with mercury to about the level TT'. We cannot describe fully the method of using the thermometer, but can only state here that by adjusting the height of the tube T' we can keep constant either the pressure or the volume of the air in the bulb, and so use it either as a constant-pressure or constant-volume thermometer; and that by means of this apparatus we can test the accuracy of the laws described in Arts. 31-33.

### EXAMPLES ON CHAPTER V

In the following examples the coefficient of expansion of air and other gases may be taken as $\frac{1}{273}$. The equations required are given in Arts. 31-34 with some solved examples; but beginners will find it well to work out the simpler questions (4-11) in the way the first two are worked out below, *i.e.* by applying Charles's law directly. Students should pay especial attention to problems on the expansion of gases, because such calculations have frequently to be made in physical and chemical work. The volumes of gases are generally measured at temperatures higher than 0° and under varying barometric pressures. In order to compare the results of experiments made under different conditions of pressure and temperature, it is necessary to find the volume which the gas would occupy under a pressure of 76 cm. of mercury and at the temperature of 0° C. This process is known as 'reducing the volume to the normal pressure and temperature,' and is illustrated in Ex. 3.

1. What change of volume will be produced by heating 26 litres of air from 0° to 21°?

According to Charles's law the increase of volume for each degree rise in temperature is $\left(26 \times \frac{1}{273}\right)$ litre. For a rise of 21° the increase is $\frac{26 \times 21}{273} = \frac{26}{13} = 2$ litres, so that the volume at 21° is $26 + 2 = 28$ litres.

2. A gas-bag contains 14 cubic feet of coal-gas at 0° and under a pressure of 26 lbs. per square foot: if the pressure is increased to 28 lbs. per square foot what will the volume become? To what temperature would the gas have to be warmed so as to recover its original volume?

When the pressure increases from 26 to 28 lbs. per square foot the volume diminishes in the inverse proportion and becomes $14 \times \frac{26}{28} = 13$ cubic feet.

We have next to find what rise of temperature above 0° would bring back the volume from 13 to 14 cubic feet. The expansion per degree is $\frac{13}{273} = \frac{1}{21}$ cubic feet. The expansion for $x$ degrees would be $\frac{x}{21}$, and this must be equal to 1 cubic foot. Hence $x = 21$, and the required temperature is 21°.

3. On heating a certain quantity of mercuric oxide it is found to give off 380 c.c. of oxygen gas, the temperature being 23° and the barometric height 74 cm.: what would be the volume of the oxygen measured at the normal pressure and temperature?

First allow for the change of pressure, according to Boyle's law, assuming the temperature to remain constant. Since the volumes are inversely proportional to the pressures, the gas would occupy under a pressure of 76 cm. a volume of $380 \times \frac{74}{76} = 370$ c.c.

Next allow for the change of temperature, the pressure remaining constant and equal to 76 cm. The absolute temperatures corresponding to 23° and 0° are 296° and 273° respectively. At the latter temperature the volume would be

$$370 \times \frac{273}{296} = \frac{2730}{8} = 341\tfrac{1}{4}.$$

Thus the volume at the normal pressure and temperature would be $341\tfrac{1}{4}$ c.c.

4. A certain quantity of gas measures 260 c.c. at 0°: what would be its volume at 63°?

5. To what temperature must a gas be heated in order that its volume may become double of what it is at 0°?

6. A certain quantity of gas measures 90 c.c. at 0°: at what temperature will its volume become 120 c.c.?

7. The volume of a gramme of hydrogen at 0° is 11·16 litres: what is its volume (1) at 30°, (2) at 50°?

8. 200 cubic centimetres of air are heated from 0° to 30° and at the latter temperature the volume is found to be 222 c.c.: what value does this give for the coefficient of expansion of air?

9. A closed glass tube filled with air at 0° and under atmospheric pressure is gradually heated. If the tube can safely stand a pressure of 3 atmospheres, to what temperature may it be heated?

10. A litre flask contains 1·293 gm. of air at 0°: how much will it contain at 100°?

11. The pressure upon a gas is doubled, and at the same time its temperature is raised from 0° to 91°: how is the volume altered?

12. A certain quantity of gas occupies a volume of 66 c.c. at 13°: what will be its volume at 52°? At what temperature will its volume be 63 c.c.?

13. At what temperature will the volume of a given mass of gas be exactly double of what it is at 30°?

14. The pressure upon a gas is doubled, and at the same time its temperature is raised from 13° to 299°: how does this affect its volume?

15. Find the volume at the normal pressure and temperature of a quantity of gas which measures 392 c.c. at 21°, and under a pressure of 80 cm. [See Ex. 3.]

16. If 3000 cubic inches of air at 0° C. expand by 11 cubic inches for each degree rise of temperature, find the volume at 100° of a quantity of air which at 50° measures 100 cubic inches, the pressure being supposed to undergo no change.

17. The pressure inside a steel cylinder containing compressed oxygen is measured by means of a manometer, and is found to be 30 atmospheres at a temperature of 27°. The cylinder is now surrounded by a freezing mixture, which reduces the temperature to $-13°$, and the pressure falls to 26 atmospheres. Find the coefficient of increase of pressure.

# CHAPTER VI

## SPECIFIC HEAT AND CALORIMETRY

**35. Distinction between Temperature and Heat.—** We now proceed to that branch of our subject which is called calorimetry, and which treats of heat as a measurable quantity. The important difference between temperature and heat may be illustrated by comparing it with the difference between the 'level' or height of a cistern and the quantity of water in that cistern. If two cisterns at different levels are connected by a tube, the water will tend to flow from the higher cistern to the one at a lower level: so in the case of two bodies in thermal communication, heat tends to flow from the one at a higher temperature to the one at a lower temperature. But you cannot tell from the height of the cistern how much water there is in it, or how much work you could make the water do by turning a wheel. Similarly, the thermometer indicates the temperature of a body, but does not of itself tell us how much heat we can get out of the body.[1] You may have a bucketful of hot water and a thimbleful of water equally hot: they are at the same temperature, but it is clear that the bucketful would give out more heat than the thimbleful in cooling through the same number of degrees. And this is equally true of the amounts of heat required to warm them through equal ranges of temperature.

[1] Avoid the use of the term 'heat contained in a body.' It is misleading; at any rate it is of no practical importance, for we are only interested in finding how much heat is given out or absorbed by a body when it is cooled or heated, and this obviously depends not only upon its temperature at the time but also upon the temperature to which it is cooled or heated.

**36. Heat as a Measurable Quantity.**—In measuring lengths we express them in terms of a certain unit of length—the *centimetre:* we state the mass of a body as being so many grammes—using the *gramme* as the unit of mass; and so, in order to measure and express quantities of heat, we require to choose a *heat-unit*, which is defined as follows:—

*The unit of heat is the amount of heat required to heat a gramme of water through one degree.* This is the same as the amount of heat given out by a gramme of water in cooling through 1°.

It is clear that the amount of heat required to raise the temperature of 2 grammes of water through 1° is twice as great as that required to heat 1 gramme through 1°—or is equal to 2 heat-units: to heat 10 grammes through 1° would require 10 units; and so on, the quantity of heat being proportional to the mass of the substance heated.

Again, experiment shows that the amount of heat required to warm a body through a given range of temperature is proportional to that range of temperature, *i.e.* to the number of degrees through which it is heated. Thus: to heat 1 gramme of water from 0° to 100° we require 100 units (the same amount as would be required to heat 100 grammes from 0° to 1°). To heat 100 grammes of water from 0° to 100° we require 100 × 100 = 10,000 units. A kilogramme (1000 grammes) of water in cooling from 75° to 10° gives out 1000 × 65 = 65,000 units of heat. In general, the amount of heat required to raise $m$ grammes of water through $\theta°$ is given by the equation

$$H = m\theta \qquad \qquad . \qquad . \qquad . \qquad (1)$$

**37. Specific Heat.**—The question now arises—If we take equal weights of *different* substances and heat them through equal intervals of temperature, shall we find any differences between the amounts of heat required? The following experiments supply the answer and show that there are such differences.

EXPT. 17.—Apparatus required—A few metal balls or bullets (with hooks attached) made of lead, iron, tin, bismuth, etc.; a cake of beeswax (or a mixture of beeswax and vaseline) about 6 mm. or quarter of an inch thick: this may be made by pouring the melted wax into a flat dish or by melting the wax on the surface of hot water; an oil-bath.

Suspend the balls from strings, or on a wire support (as in Fig. 21), and heat them to about 150° in the oil-bath. Drop them simultaneously on to the wax cake. The iron ball melts through first, and is followed by copper, zinc, and tin. Lead is barely able to get through: the bismuth ball will probably not get through at all.

Now the rate at which any ball melts through depends chiefly upon the amount of heat which it gives out in cooling, for this determines how much wax it will melt; it also depends upon the density (Art. 24) of the ball, for the densest ball would tend to sink fastest through the melting wax, as it would if the balls were thrown into treacle. The lead ball is heavier than any of the others: the density of lead is 11·4, while that of iron is 7·2. Although the lead is so much heavier it does not get through nearly as quickly as the iron. We conclude from this that iron in cooling gives out more heat than lead does.

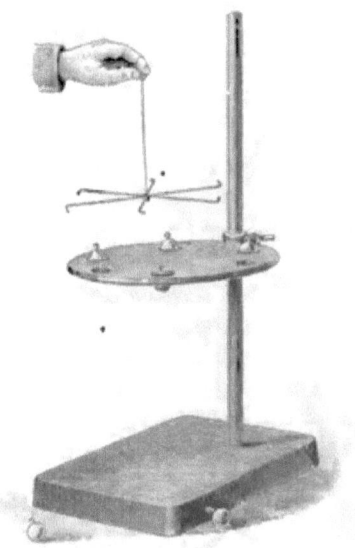

Fig. 21 [1/8].

EXPT. 18.—Place equal quantities (about 400 c.c.) of water in two beakers, and in these immerse the two bulbs of the differential thermometer (Art. 34). Take equal weights (100 grammes) of water and any metal, say lead shot, and put them in separate test-tubes. Plunge the two test-tubes into the same vessel of hot water, and, after keeping them there for 10 minutes, pour the hot lead into the left-hand beaker and the hot water into the other, moving the bulbs of the thermometer slightly so as to stir the contents of the beakers. The index of the differential thermometer will move, showing that the contents of the right-hand beaker have been warmed more than those of the other.

We conclude from this that the water gives out more heat than the lead in cooling through the same interval of temperature,[1] and this is found to hold for all other metals. The experiment may be varied by immersing equal weights of different metals in the two beakers: it will be found that similar differences exist among the metals. This is expressed by saying that every substance has a distinct **specific heat,** and this term is defined as follows:—

*The specific heat of a substance is the ratio between the amount of heat required to raise a given mass of that substance through a given interval of temperature and the amount of heat required to raise an equal mass of water through the same interval of temperature.*

Thus when we say that the specific heat of mercury is $\frac{1}{30}$ or 0·033, we mean that the amount of heat required to raise a given mass of mercury through any interval of temperature is only one-thirtieth of the amount of heat required to raise an equal mass of water through the same interval of temperature; or that in cooling through the same range of temperature water would evolve thirty times as much heat as an equal mass of mercury.

We have seen (Art. 36) that $m\theta$ units of heat are required to cause a rise of temperature of $\theta°$ in $m$ grammes of water (the specific heat of which is unity). For any other substance, the specific heat of which is less than unity and equal to $s$ say, the amount of heat required will be less in the same proportion and will be given by the equation

$$H = ms\theta \qquad . \qquad . \qquad . \qquad . \qquad (2)$$

which also expresses the amount of heat given out by a body of mass $m$ and specific heat $s$ in cooling through $\theta°$.

The following examples will illustrate the use of this important equation:—

*Ex.* 1.—How much heat is required to raise the temperature of a kilogramme of mercury (specific heat = 0·033) from 20° to 170°?

Here $\theta = 170° - 20° = 150°$, and $m = 1$ kgm. $= 1000$ gm.

$\therefore H = ms\theta = 1000 \times 0·033 \times 150 = 4950$,

and the amount of heat required is 4950 units.

---

[1] The fact that the hot water really falls through a somewhat smaller range of temperature only makes the experiment more conclusive.

*Ex.* 2.—If 342 units of heat are imparted to 150 grammes of iron (specific heat = 0·114) originally at 10°, what will be the final temperature of the iron?

If we suppose the temperature of the iron to be raised through $\theta°$, we know that

$$342 = 150 \times 0\cdot 114 \times \theta,$$

and

$$\therefore \theta = \frac{342}{150 \times 0\cdot 114} = \frac{342}{17\cdot 1} = 20°.$$

Thus the final temperature of the iron will be $10° + 20° = 30°$.

**38. Methods of Finding Specific Heats.**—(1) The Method of Mixtures.—The specific heat of a substance may be found by heating it to a known temperature and dropping it into a weighed quantity of water contained in a suitable vessel (called a calorimeter). The temperature of the water is observed by means of a thermometer before and after dropping in the hot substance, and from the rise of temperature we can calculate the amount of heat gained by the water. This must be exactly equal to the amount of heat given out by the hot body, provided that care is taken to prevent loss or gain of heat from surrounding bodies during the experiment. Having thus found how much heat the substance gives out in cooling through a known interval of temperature, we can easily calculate its specific heat. This method is described more fully below (Expt. 19).

(2) The Method of Cooling.—The specific heat of a substance can also be found by heating it and allowing it to cool; the rate at which it cools is observed and compared with the rate at which water cools under similar conditions. For example, put equal weights of water and turpentine, both at 50°, into two similar glass beakers or thin metal vessels (suspended by strings so as not to be influenced by surrounding bodies), and put a thermometer in each. Note the time taken by each to cool from 50° to 20°; you will find that the water takes more than twice as long as the turpentine. Since the turpentine cools under just the same conditions as the water, the difference must be due to its lower specific heat, which is less than half that of water.

(3) The Method by Fusion of Ice.—The amount of heat given out by a hot body in cooling to 0° can also be measured by finding how much ice it is able to melt. This method of

determining specific heats will be described in the next chapter.

EXPT. 19.—*To find the specific heat of lead shot by the method of mixtures.* Apparatus required—A steam-heater: this can be made by fitting a wide test-tube *loosely* into the neck of a corked flask; a bent tube should also be fitted into the cork so as to allow the steam to escape (Fig. 22). A thermometer. A beaker holding about 200 c.c. to serve as a calorimeter: a cylindrical vessel made of thin sheet-brass does better as a calorimeter because it absorbs less heat; a small tin pot, such as can be obtained at a grocer's or chemist's shop does fairly well.

Fig. 22.

Weigh into the calorimeter 100 grammes of water (or as much as will half fill it), and place the calorimeter on a flat cork or wrap it round with wool or flannel (to prevent loss of heat by conduction). Put 200 grammes of lead shot into the test-tube and close the mouth of the tube with a plug of cotton-wool to keep out cold air. Boil the water in the flask, and keep it boiling for quarter of an hour so that the lead may get heated to the temperature of the steam. When all is ready, observe carefully by means of the thermometer the temperature of the water in the calorimeter; suppose this to be $9°·6$. Now remove the cotton-wool, lift the test-tube out and quickly pour the shot into the calorimeter. Stir the water gently by moving the bulb of the thermometer, and note the highest temperature which it reaches. Suppose this to be $14°·9$; we can now calculate the specific heat of the lead shot.

Let $s$ be the required **specific heat**. The lead has cooled from 100° to 14°·9, *i.e.* through 85°·1. By equation (2) the amount of heat it has given out in cooling is

$$H = 200 \times s \times 85 \cdot 1 \text{ units.}$$

This has been absorbed by the water, and has raised its temperature from 9°·6 to 14°·9, *i.e.* through 5°·3. The amount of heat required for this is

$$H = 100 \times 5 \cdot 3.$$

Since the two quantities are equal, **we have**

$$200 \times 85 \cdot 1 s = 100 \times 5 \cdot 3,$$

and

$$\therefore s = \frac{5 \cdot 3}{170 \cdot 2} = 0 \cdot 031.$$

Make similar experiments with **mercury**, iron **(nails)**, copper (short pieces of thick wire), etc.

**39. Thermal Capacity.**—The specific heat of a substance depends only upon the nature of the substance: it does not depend upon its size or weight, any more than its hardness or colour does. The specific heat of a brass button is the same as that of a brass cannon—if both are made of the same kind of brass. But the quantity of heat required to warm a body through a given range of temperature depends not only upon its specific heat ($s$) but also upon its mass ($m$). The product ($ms$) of these two quantities is called the *thermal capacity* or *capacity for heat* of the body, and may be defined as follows :—

*The thermal capacity of a body is measured by the amount of heat required to raise its temperature by one degree.*

*Ex.* 3.—What is the thermal capacity of a calorimeter weighing **100** gm. and made of copper of specific heat 0·095?

Since 0·095 is the specific heat of copper, **0·095** of a heat-unit is required to raise 1 gm. of copper 1°. The amount of heat required to raise 100 gm. of copper 1° is $100 \times 0.095 = $ **9·5** units. Thus the thermal capacity of the calorimeter **is** 9·5.

**40. Results and Applications.**—In the following table the approximate values of the specific heats of a few substances are given :—

| | | | | | | | | |
|---|---|---|---|---|---|---|---|---|
| Ice | . | . | . | **0·5** | Water . | . | . | 1·000 |
| Glass | . | . | . | **0·2** | Alcohol | . | . | 0·615 |
| Sulphur | . | . | . | 0·18 | Turpentine . | . | . | 0·425 |
| Iron . | . | . | . | 0·114 | Mercury . | . | . | 0·033 |
| Copper | . | . | . | 0·095 | Air (at constant pressure) | | | |
| Lead | . | . | . | **0·031** | sure) | . | . | 0·237 |

Observe that mercury has a very low specific heat. This is another reason in favour of its use in thermometers. When we wish to find the temperature of a hot liquid and dip a thermometer into it, the bulb and its contents absorb some heat, and so the liquid is slightly cooled : but the amount of heat so absorbed by a mercurial thermometer is much less than it would be, for example, if the bulb were filled with water.

Water has a higher specific heat than any other liquid or solid. For this and other reasons the sea is not warmed to nearly the same extent as land by the sun's rays; nor does it cool so rapidly when they are absent. As the air above it partakes of its temperature, it follows that islands and places on or near the sea-coast enjoy a more equable climate than inland places. The differences between summer and winter temperatures in our island are only about half as great as the corresponding differences in regions of central Russia which are in the same latitude.

On account of its high specific heat, water not only cools slowly but gives out a large amount of heat in the process. Among familiar applications of this we may mention hot water bottles, the foot-warmers used in railway carriages, and the systems of heating by hot-water pipes commonly used in greenhouses and public buildings.

### Examples on Chapter VI

Read Arts. 37-39 again before you attempt to work these out. The specific heats required may be taken from the table in Art. 40. You will generally find that the best and simplest method of solving a specific heat problem is the method we adopted in working out the results of Expt. 19. Find the amounts of heat given out by the hot substance and taken in by the cold substance : if no heat is lost or gained in the experiment, these must be equal and can be equated.

1. A certain vessel holds 800 c.c. of water at its temperature of maximum density ($4°$). How much heat must be imparted to the water before it begins to boil?

2. How many units of heat are required to raise the temperature of 150 gm. of copper from $10°$ to $150°$?

3. What is the thermal capacity of a leaden bullet weighing 100 gm.?

4. Find the specific heat of a substance 125 gm. of which at $78°$, when immersed in 250 gm. of water at $12°$, gave a resulting temperature of $18°$.

5. Two pounds of boiling water are poured upon 10 lbs. of mercury at 16°: what will be the common temperature after mixing?

6. What is meant by saying that the 'specific heat' of water is thirty times as great as that of mercury? If a pound of boiling water be mixed with 3 lbs. of ice-cold mercury, what will be the temperature of the mixture?

7. If a kilogramme of mercury at 120° is poured into a vessel containing 200 gm. of ice-cold water, what will be the temperature after the whole is mixed? How would the weight and material of the vessel affect the result?

8. How is the climate of the British Isles affected by the high capacity for heat of water?

9. 30 grammes of iron nails at 100° are dropped into 60 gm. of water at 13°·2, and the final temperature is found to be 18°·6: what is the specific heat of the nails?

10. What is meant by the statement that the specific heat of platinum is 0·03?

In order to ascertain the temperature of a furnace, a platinum ball weighing 80 grammes is introduced into it: when this has attained the temperature of the furnace it is quickly transferred to a vessel containing 400 grammes of water at 15°. The temperature of the water rises to 20°: what was the temperature of the furnace?

# CHAPTER VII

## CHANGE OF STATE—FUSION AND SOLIDIFICATION

**41. Fusion.**—We have seen that the application of heat to a solid causes a rise of temperature and an increase in size: in most cases, when the temperature is raised sufficiently, the solid is converted into a liquid. This change of state is called melting or fusion. Thus ice melts at 0° to form water: sulphur melts to a yellow liquid when heated somewhat above the boiling-point of water. Crystals of iodine placed in a flask and gently heated over a flame melt to an almost black liquid; this on further heating undergoes a further change of state and boils, filling the flask with a splendid purple vapour.

There are some substances, such as carbon and lime, which have not yet been fused because we are not able to heat them to a sufficiently high temperature. Others, such as iron, glass, sealing-wax, and pitch, before they become liquid pass through an intermediate stage in which they are plastic and can be easily moulded into any shape. It is in this intermediate stage that the glass-blower works glass before the blowpipe and the smith works and welds wrought-iron.

**42. Melting-Points.**—But, in general, the change from the solid to the liquid state is well marked and occurs at a definite temperature. This is called the temperature of fusion or the **melting-point** of the substance. The reverse change—solidification—takes place at the same temperature, provided the liquid is stirred or shaken as it is cooled; and thus we speak of 0° C. either as the melting-point of ice or the freezing-point of water. As this is taken as one of the standard points in measuring temperature, you should convince yourself by some such experiments as the following that it is really a fixed point.

EXPT. 20.—Pound some clean ice and put it in a funnel: pour on it some water to wash it, and then immerse a thermometer in it as in Art. 7. Read off the height of the mercury.

Put the ice into a beaker and pour water on it. Stir both well round with the thermometer: observe that it indicates the same temperature as when put into melting ice, and that this temperature does not depend upon the proportion of ice and water in the mixture.

Repeat the experiment with hot water. This melts some of the ice, but is itself cooled thereby. Warm the mixture with a lamp. Remove the lamp and stir the ice and water well up. As long as any ice is left the temperature does not rise above 0°.

EXPT. 21.—Add a little common salt to the ice and water in the preceding experiment. The thermometer will fall below 0°. The salt lowers the melting-point (see Art. 45). Clean ice should therefore be used in marking the lower fixed point of a thermometer.

EXPT. 22.—To find the melting-point of beeswax. Draw out a piece of glass tubing (or a test-tube) before the blowpipe into a fine tube. Cut off a piece about 2 inches long and seal up one end in the flame. Introduce a few small pieces of the wax. Tie the tube on to a thermometer, so that the wax is near the bulb, and dip it into a beaker of water. Heat gradually, stirring the water, and observe the temperature at which the first drop runs down the tube. This is the melting-point.

TABLE OF MELTING-POINTS.

| Mercury | . | . | . | $-38°\cdot5$ | Bismuth | . | . | . | $265°$ |
|---|---|---|---|---|---|---|---|---|---|
| Ice | . | . | . | $0°$ | Cadmium | . | . | . | $320°$ |
| Beeswax | . | . | . | $65°$ | Lead | . | . | . | $330°$ |
| Sulphur | . | . | . | $115°$ | Cast-iron | . | . | . | $1200°$ |
| Tin | . | . | . | $230°$ | | | | | |

Mixtures and alloys generally melt at lower temperatures than their constituents. Thus an alloy of bismuth, lead, tin, and cadmium, known as Rose's fusible metal, melts at about 70°. The melting-points of its constituents are all above 200°: but a slip of the alloy placed in boiling water fuses readily.

**43. Heat Absorbed in Melting.**—When heat is applied to a mixture of ice and water the ice gradually melts, but the temperature remains at 0° provided the mixture is constantly stirred. As the heat does not produce any rise of temperature it must be spent in converting the ice into water. The heat thus spent in producing change of state without change of temperature is called the **latent heat of fusion**, and may be defined as follows:—

E

The latent *heat* of fusion of ice (*or the latent heat of water*) is the amount *of heat required* to convert unit mass (*one gramme*) of ice at 0° *into* water at the same temperature.

The latent heat of water is 80: by saying this we mean that 80 units of heat are required to melt 1 gramme of ice at 0° without raising its temperature.

Think over this in connection with frosts and thaws. Try to realise that before a single pound of ice can be melted, as much heat must be imparted to it as would raise 80 lbs. of water through 1°, or 1 lb. of water from 0° to 80°; and that before a single pound of ice can be frozen, the same amount of heat must be taken away from it. This will help you to understand how roads remain covered with half-frozen slush long after a thaw has set in; and how it takes several frosty nights to freeze over a pond or lake (see also Art. 25).

EXPT. 23.—*To find the latent heat of water.* This can be found by putting a known quantity of ice into warm water and observing how much heat is taken from the water to melt the ice.

Weigh out 150 grammes of water at about 40° into a beaker or calorimeter (such as was used in Expt. 19). Note the total weight of calorimeter and water. Weigh out (roughly) about 60 grammes of pounded ice. Dry it quickly, first with a towel and then with blotting-paper. After taking the temperature of the water, pour in the ice: stir it round with the thermometer (or a wire stirrer) and observe the lowest temperature reached when all the ice is melted. Now weigh the calorimeter and its contents again; the increase of weight tells us how much ice has been put in. We shall suppose that this is 50 grammes, and that the final temperature of the water is 10°.

Now let $x$ be the latent heat of fusion that we wish to find; *i.e.* the number of heat-units required to melt 1 gramme of ice. Clearly $50\,x$ are required to melt 50 grammes. Further, the 50 grammes of ice-cold water thus produced have been warmed up from 0° to 10°: this requires a further supply of $50 \times 10 = 500$ units.

The whole of the heat thus absorbed has been supplied by 150 grammes of water in cooling from 40° to 10°, in which

process we know that $150 \times 30 = 4500$ heat-units are given out
Thus $50x + 500 = 4500$,
$\therefore 50x = 4000$ and $x = 80$.

Thus the latent heat as determined by our experiment is 80.

**44. Ice-Calorimeters.**—You will now be able to understand how it is possible to determine the specific heat of a substance by finding out how much ice it can melt in cooling down (say) from 100° to the temperature of melting ice (0°). For when we know this we know how much heat the substance has given out—every gramme of ice melted corresponding to 80 units of heat evolved.

Black (who discovered the law above referred to) used an

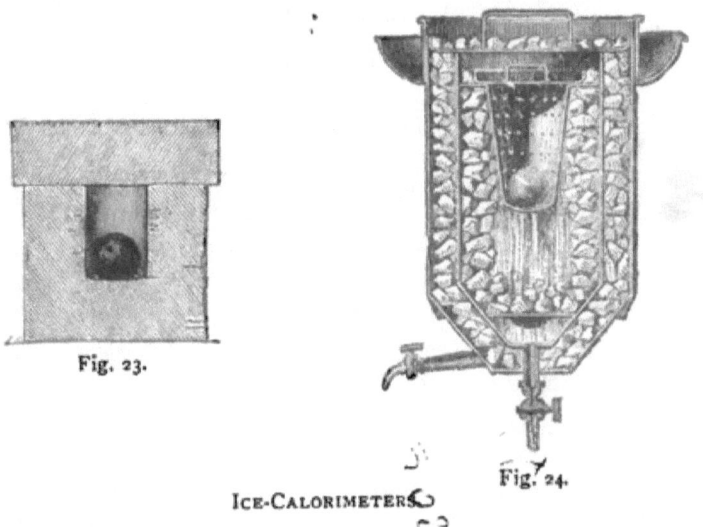

Fig. 23.   Fig. 24.
ICE-CALORIMETERS

ice-calorimeter of the very simple form shown in Fig. 23. The hot body was dropped into a hole scooped out of a block of ice, which was immediately covered with another slab of ice. Lavoisier and Laplace improved the ice-calorimeter and gave it the form shown in Fig. 24. The hot body is dropped into the inner vessel, which is made of perforated metal: the heat which it gives out in cooling melts some of the ice in the middle vessel, and the water thus produced is run off by the stopcock at the bottom and measured. The outer vessel, also filled with ice, simply serves as a screen to protect the middle vessel from the surrounding atmosphere, which would otherwise gradually melt the ice in it. The way in which the results are worked out will be understood from the following example.

A copper ball weighing 600 grammes was heated to 100° and introduced into an ice-calorimeter. The water produced was found to weigh 72 grammes. Find the specific heat of copper.

Let $s$ be the required specific heat. By equation (2), Art. 37, the amount of heat given out by the copper in cooling from 100° to 0° is $ms\theta = 600 \times s \times 100$. This was entirely spent in melting 72 grammes of ice, for which purpose we know that $72 \times 80 = 5760$ heat-units are required. These two quantities are therefore equal, i.e.

$$60,000\, s = 5760,$$

and
$$\therefore s = \frac{576}{6000} = 0.096.$$

**45. Freezing Mixtures.**—EXPT. 24. Dissolve a spoonful of common salt or of sodium sulphate in a cupful of water, stirring it up with a thermometer. Observe that the water becomes colder as the salt dissolves. This is generally the case: solution, like liquefaction, requires a supply of heat, and if the heat required is not supplied from the outside, it is taken from the materials themselves, which are thereby cooled. Ammonium nitrate shows this cooling effect better than either of the substances named.

Freezing mixtures, such as are used for the artificial production of ice, depend upon this principle. If ammonium nitrate at 0° is mixed with an equal weight of water at the same temperature, the solution will be found to assume the low temperature of $-15°$. Another common freezing mixture (used by confectioners in making 'ices') consists of equal weights of common salt and pounded ice or snow. In cases like this, where both materials are solid, there is a double absorption of heat, for (1) heat is absorbed in melting the ice, and (2) a further supply of heat is required to make the salt dissolve.

EXPT. 25.—Fill a metal vessel (tin pot or saucepan, or brass calorimeter) with either of the above-mentioned freezing mixtures and place it on a flat saucer, into which a *little* water has been poured. Stir the mixture with a stick. In a few minutes the water will have frozen, and you will be able to lift up the pot with the saucer frozen to it.

Pour a little water into a test-tube and plunge it into the mixture, stirring it about from time to time. In about half an hour the water will be converted into a cylinder of ice. Notice that the outside of the vessel soon gets covered with dew and later on with ice (hoar-frost).

The common practice of sprinkling salt over pavements to clear off snow is in many ways bad. It converts the snow into a slushy freezing-mixture: and even if this is swept and washed off, it leaves the pavement so cold that the water freezes on it into a smooth and slippery sheet of ice.

**46. Expansion of Water on Freezing.**—Most substances contract on solidifying, but water is a remarkable exception to this rule. It expands on freezing, 10 c.c. of water forming 11 c.c. of ice. Consequently ice is lighter than water and floats upon its surface.

EXPT. 26.—Fit a small flask with a cork and glass tube as in Fig. 25. Put into the flask a handful of pounded ice and fill it with coloured water. Fit the cork and tube into the neck, pushing the cork down until the water rises to the level of the mark on the card. Put the flask in a basin of luke-warm water. Observe the contraction that takes place as the ice melts.

EXPT. 27.—Blow a small bulb at one end of a glass tube and draw the other end out to a fine point. Fill the tube with water and seal off the point in a blowpipe-flame. Place the bulb in a freezing mixture and cover it with a cloth. The bulb will soon burst, owing to the expansion of the water in freezing.

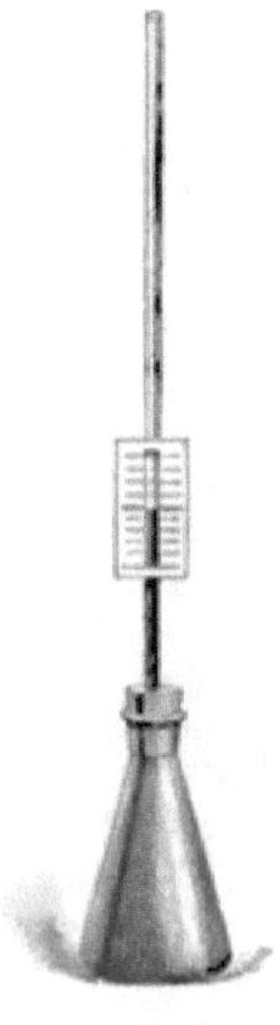

Water expands in freezing with almost irresistible force. Not only thick glass bottles, such as soda-water bottles, but even cast-iron shells half an inch thick are easily burst by filling them with water and exposing them to the frost. Thus water-pipes often burst during frosty weather, as householders find out to their disgust when the thaw sets in and their houses are flooded with water. It is usual to cover water-pipes and

Fig. 25 [1/10].

54        HEAT        CH.

hydrants with straw and other non-conducting substances to protect them from the action of frost; and, as running water does not readily freeze, taps are often kept open (in spite of municipal regulations!) as long as the frost lasts.

### Examples on Chapter VII

Read Arts. 43, 44 again; also the note at end of Chap. VI. Whenever weights are expressed in pounds (instead of grammes) use as your unit of heat the 'pound-degree' or amount of heat required to raise the temperature of 1 lb. of water through 1°.

The latent heat of water may be taken as 80.

1. What will be the result of mixing 10 lbs. of snow at 0° with 4 lbs. of water at 60°?

    First find whether all or only part of the snow will be melted. The amount of heat required to melt 10 lbs. of snow would be $10 \times 80 = 800$ pound-degree units of heat. The temperature of the mixture cannot fall below 0°, and 4 lbs. of water in cooling from 60° to 0° only give out $4 \times 60 = 240$ pound-degrees, or enough to melt $\frac{240}{80} = 3$ lbs. of snow. Thus the result will be a mixture of 7 lbs. of snow and 7 lbs. of water, all at 0°.

2. How much ice at 0° will be melted by 1000 gm. of boiling water?

3. When equal weights of boiling water and melting ice are mixed, the ice all melts, and the resulting temperature is 10°: find from this the latent heat of fusion.

4. How much hot water at 75° will just melt 30 lbs. of ice?

5. 300 grammes of melting ice are mixed with 700 grammes of boiling water, and the resulting temperature is 46°: what is the latent heat of fusion?

6. A brass cylinder weighing 80 gm. was heated to 100° and dropped into an ice-calorimeter. The amount of ice melted was 9 gm.: find the specific heat of the brass.

7. Two copper balls of the same weight and raised to the same temperature are laid, the one on a cake of slowly-melting ice, and the other on a cake of wax. The latter sinks in the more deeply  What inference would you draw from this?

8. Explain the statement that the latent heat of water is 80. To a pound of ice at 0° are communicated 100 units of heat (pound-degrees Centigrade). What change of temperature does the ice undergo, and in what way is its volume altered?

9. 143 grammes of water were obtained when a kilogramme of iron at 100° was introduced into an ice-calorimeter: what was the specific heat of the iron?

10. A quantity of ice is thrown into a basin containing 4 lbs. of water

at 30°, and after all the ice has melted the temperature is found to have fallen to 8°: how much ice was thrown in?

11. What is meant by a *unit of heat?* Taking the specific heat of lead as 0·031, and its latent heat as 5·07, find the amount of heat necessary to raise 15 lbs. of lead from a temperature of 115° C. to its melting-point, 325° C., and to melt it.

# CHAPTER VIII

## CHANGE OF STATE—VAPORISATION AND CONDENSATION

**47. Evaporation and Boiling.**—There are two ways in which a liquid may be changed into a vapour. The first of these is the slow process which is called **evaporation**. You have doubtless noticed that, if water is left exposed in a shallow dish or saucer, it gradually dries up or evaporates. This evaporation of water goes on at all ordinary temperatures, but most rapidly in warm dry weather. Alcohol (methylated spirit) evaporates more rapidly than water; and ether evaporates so quickly that, if you pour a few drops on the palm of your hand, the ether will disappear in a few seconds, and produce an intense sensation of cold.

The second process is the familiar one of **boiling**. You should begin your study of this by boiling some water in a glass flask. When the cold water is heated, the first thing you notice is the expulsion of the air which it always contains in solution. This makes its appearance in the form of minute bubbles, which gradually rise through the water and are expelled. Larger bubbles, of steam or water-vapour, soon make their appearance, but as they rise into the upper and cooler water, they condense and collapse with a peculiar rattling noise. This is what causes the 'singing' of a kettle —always a sign that boiling is not far off. Finally bubbles of steam begin to rise rapidly and freely from all parts of the water, and the actual boiling sets in as soon as the whole of the liquid has been heated to a certain temperature, called its **boiling-point**. Observe that the steam itself is, like air, invisible; the cloud that forms above the mouth of the flask consists of condensed water-particles, not water-vapour.

The distinction between the two processes (evaporation and boiling) is this. Evaporation goes on at *all* temperatures, but only from the *surface* of the liquid. Boiling consists in the rapid production of bubbles of vapour *throughout the mass of the liquid*, and only takes place *at a definite temperature* (the boiling-point of the liquid).

EXPT. 28.—*To find the boiling-point of a liquid.* All that is needed for a rough determination is a small long-necked flask or test-tube (Fig. 26), fitted with a double-bored cork, in which are inserted the thermometer and a bent tube for the escape of the vapour. The thermometer should be pushed down until the bulb nearly touches the liquid, but it should not dip into it. Alcohol or methylated spirit will do well for the experiment. If you wish to recover the liquid, you can do so by connecting the bent tube to a condenser (see Art. 52). If the liquid boils by fits and starts, or 'bumps,' you should put into the test-tube a few bits of crumpled tin-foil or platinum-foil. This will make the boiling more regular.

Fig. 26.

48. **Vapour-Pressure.**— When water is introduced into a closed space containing air (*e.g.* a corked flask), it goes on evaporating slowly until the air contains as much water-vapour as it can hold in suspension; the evaporation then stops, and the air is said to be *saturated with water-vapour*.

If the water is introduced into a vacuous space (*e.g.* a flask out of which all the air has been pumped), the evaporation goes on more rapidly; practically it is instantaneous. You already know that the vapour of boiling water (steam) exerts a pressure: it is by this pressure of steam that all steam-engines work. You have now to learn that even at ordinary temperatures water-vapour exerts a small but measurable pressure. This pressure can be observed and measured by introducing water into a Torricellian vacuum (Art. 28).

EXPT. 29.—Fill a barometer-tube with mercury and invert it as in Expt. 15. Make a bent pipette, fill it with the liquid to be introduced, and push the curved point of it under the mercury and into the tube (Fig. 27). Ether is the best liquid to start with. Blow *cautiously* into the upper end of the pipette, so as to make the ether ascend *drop by drop*, and watch the effect. The first drop will evaporate almost before it reaches the vacuum, and the column of mercury

Fig. 27.

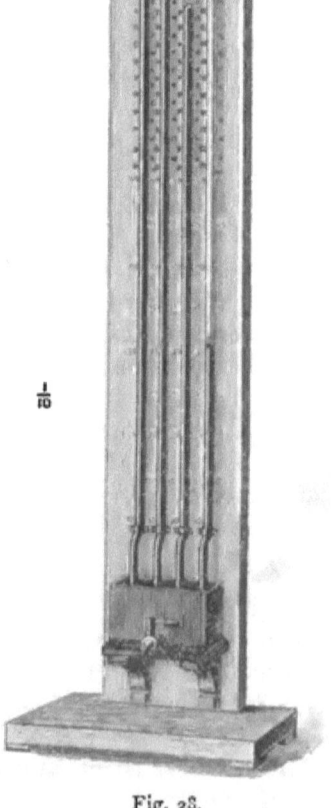

Fig. 28.

will fall through a few centimetres. This fall is due to the pressure exerted by the ether-vapour, for there is nothing else above the mercury column to exert any pressure. As the ether ascends drop by drop, the mercury falls lower and lower until a point is reached, after which the introduction of more ether does not cause any further depression: the evaporation stops, and a layer of liquid ether is seen resting on the mercury. Before this, the space above the mercury was only partially saturated with ether-vapour, or contained *unsaturated vapour;* it is now *saturated*, and the vapour exerts its *maximum pressure, i.e.* the greatest pressure that it can exert at the temperature of the room.

Now warm the tube cautiously by passing the flame of a spirit-lamp along it: the mercury falls lower and lower, showing that the maximum vapour-pressure *increases with the temperature*.

Perform the same experiments with alcohol and water, cleaning the tube and pipette each time, or using fresh ones. You will find that alcohol gives a much smaller depression, and water a smaller one still. Thus the vapour-pressure of a liquid *depends upon the nature of that liquid.*

In Fig. 28 are shown four tubes dipping into a mercury trough, the first tube being a reference-tube containing mercury only (a barometer, in fact), and the others containing water, alcohol, and ether respectively above the mercury. The maximum vapour-pressure in each case is measured by the depression produced, and is expressed as being equal to so many 'centimetres of mercury.' At 15° it is 1·3 cm. for water, 3·3 cm. for alcohol, and 55·4 cm. for ether.

**49. Saturated and Unsaturated Vapours.**—We may here point out the chief distinctions between the two states in which the vapour of a liquid may exist.

When a vapour cannot exist without change in the presence of its own liquid, it is called an **unsaturated vapour**. Such a vapour behaves much like a gas, and if its volume be diminished, its pressure increases according to Boyle's law (Art. 29).

On the other hand, a vapour which *can* exist without change in presence of its own liquid is called a **saturated vapour**. The pressure exerted by the vapour in this state is called the maximum vapour-pressure of the liquid, and is the greatest pressure which can be exerted by the vapour of the liquid at that temperature. It does not depend upon the volume occupied by the vapour. Thus in Fig. 28 it is not necessary that the tubes should be of the same length: if they were of different lengths, the depressions would be exactly the same. If the trough were deeper, so that any tube could be depressed in it or raised up at will, the level of the mercury inside the tube would remain unaltered.

Or, again, suppose a quantity of water and water-vapour (steam) to be contained in a cylinder closed by a movable piston. Suppose further that the temperature of the whole is kept constant, and that no air is enclosed (the space between the water and the piston being filled with saturated water-vapour). If now the piston be pushed down, the pressure will not increase (as it would in the case of a gas), but some of the vapour will condense to water, the pressure remaining constant. If the volume be increased by pulling out the piston, more of the water will evaporate to fill up the space, and the pressure will still remain unchanged. A saturated vapour, then, does not obey Boyle's law.

We may sum up briefly by saying that a saturated vapour is a vapour which is *in contact with an excess of its own liquid*. *Its pressure depends only upon* (1) *the nature of the liquid, and* (2) *the temperature.*

In what follows, unless the contrary is stated, the word vapour-pressure will be used as meaning maximum vapour-pressure or pressure of saturated vapour.

**50. Vapour-Pressure and Temperature.**—We have seen (Expt. 29) that vapour-pressure increases with temperature. We shall now show that at the boiling-point of a liquid *its vapour-pressure becomes equal to the pressure of the atmosphere.*

EXPT. 30.—Fill and invert a barometer-tube as in the last experiment, and pass up about 1 c.c. of water into it. Surround it with a 'steam-jacket,' *i.e.* a wider tube through which a rapid current of steam is passed (Fig. 29). The mercury column will gradually fall as the tube warms up, until finally, after the steam has passed freely for a few minutes, it will have fallen just to the level of the mercury in the trough outside. Clearly when this is the case the pressure inside and outside are equal. But the pressure inside is simply due to the vapour of water at its boiling-point; and the pressure outside is that of the atmosphere.

This result enables us to define the boiling-point of a liquid as follows. *The boiling-point of a liquid is that temperature at which its vapour-pressure is equal to the pressure of the atmosphere.*

The maximum pressure of water-vapour at various temperatures is given in the following table, the pressure being expressed in centimetres of mercury.

| TEMPERATURE. | PRESSURE. | TEMPERATURE. | PRESSURE. |
|---|---|---|---|
| 0° | 0·46 cm. | 60° | 14·9 cm. |
| 10° | 0·91 ,, | 70° | 23·3 ,, |
| 20° | 1·74 ,, | 80° | 35·5 ,, |
| 30° | 3·15 ,, | 90° | 52·5 ,, |
| 40° | 5·49 ,, | | |
| 50° | 9·20 ,, | 100° | 76·0 ,, |

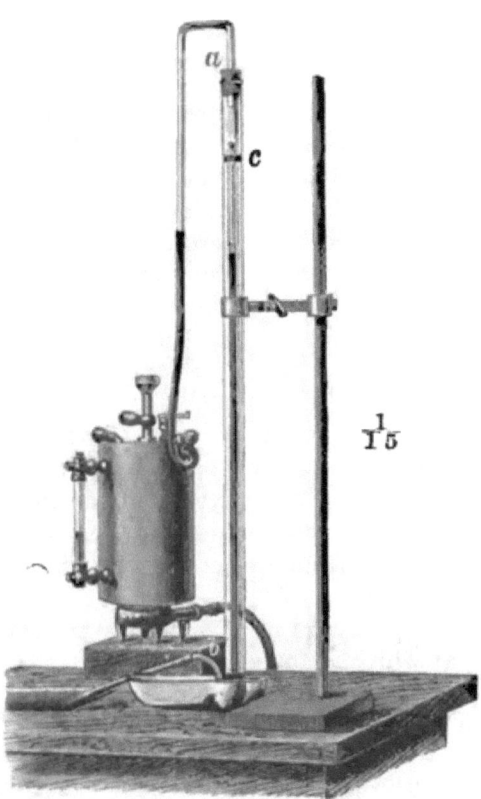

In Fig. 29 the outer tube is corked at top and bottom. Through the upper cork steam is led by the tube *a* from a flask or boiler to the steam-jacket. (An oil-can makes a capital boiler for such purposes.) The steam escapes by the pipe *b*, which passes through the lower cork and dips under the mercury. The inner glass tube should be steadied by a loosely-fitting cork at *c*.

Fig. 29.

**51. Influence of Pressure on Boiling-point.**—If the boiling-point of water be determined with a delicate thermometer from time to time, it is found to vary slightly as the atmospheric pressure varies. The experiment in the preceding article gives us the key to the connection between pressure and boiling-point. *Before a liquid can boil it must be heated to such a temperature that the pressure of its vapour is equal to the atmospheric pressure.* Thus water boils at 100° when the atmospheric pressure is equal to that of a column of mercury 76 cm. high. If the pressure increases to 77 cm., the water must be heated to a *higher* temperature before its pressure becomes equal to that of the atmosphere, and so the boiling-point is raised to 100°·34. If the atmospheric pressure is below the normal pressure, the boiling-point is lowered:

thus water boils at 99°·64 under a pressure of 75 cm. You will now understand why it is necessary to read the barometer when marking or testing the boiling-point of a thermometer.

If you gradually ascend from the sea-level up to, say, the top of a mountain, you leave more and more of the atmosphere below you, and consequently the pressure gradually becomes less. If you took with you a portable boiling-point apparatus, you would find the boiling-point become lower and lower. Thus while water boils at 100° at the sea-level, its boiling-point is 96°·3 at the top of Snowdon (3571 feet above sea-level) and 85° at the top of Mont Blanc (15,781 ft.) In Quito, which is the highest city in the world (9520 ft.), the mean height of the barometer is about 52·5 cm., so that the water boils at 90°. The change of boiling-point in ascending a mountain is one of the ways (though not the best) by which its height can be found.

The cooking and solving powers of 'boiling water' depend upon the temperature at which it boils: hence it is scarcely possible to make good tea at the top of a high mountain. By increasing the pressure we can make water boil at temperatures much higher than 100°, and so can increase its solvent power. Thus gelatine is extracted from bones by boiling them with water under great pressure in a strong closed iron boiler called a 'digester.'

**52. Condensation and Distillation.**—When steam is cooled below the boiling-point it is converted into water or condenses. If a vessel in which water is kept boiling is connected with another vessel which is always kept cool, the steam from the boiling water passes over into the colder vessel and there condenses. This process is known as distillation, and is used as a means of separating water and other liquids from impurities which they may contain in solution. Thus pure water can be obtained from sea-water by distillation, the salt and other soluble matters remaining behind in the still. Distilled water is not very pleasant to drink: it tastes 'flat,' because it contains but little air in solution.

The usual form of distillation apparatus used in laboratories is shown in Fig. 30. The liquid to be distilled is placed in a glass retort, which is heated by a burner; the neck of the retort fits into the **condenser**, which consists of a long glass tube surrounded by a wider (but shorter) tube corked at both ends. Between these two a current of cold water is kept flowing,

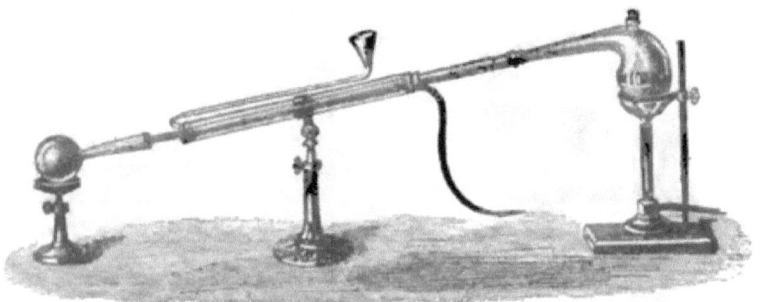

Fig. 30.

so as to cool and condense the vapour; from the condenser the distilled liquid flows into a flask or other suitable receiver.

**53. Boiling under Diminished Pressure.**—Referring to the table in Art. 50, we see that the maximum vapour-pressure of water at 90° is 52.5 cm. This is equivalent to saying that under a pressure of 52.5 cm. water would boil at 90°. Suppose, then, that you had water at 90° in a flask, and that by any means you reduced the pressure below 52.5 cm., the water would now be *above* its boiling-point, and it ought to boil without any application of heat; and this ought to occur at *any* temperature if you reduced the pressure in the flask below the maximum pressure of water-vapour at that temperature. The following experiments show that this reasoning is correct.

EXPT. 31 (Franklin's Experiment).—Half fill a flask with water, and boil it over a naked flame so as to drive out the air and fill the upper part of the flask with steam. While the water is boiling freely, remove the burner, and at the same instant close the neck of the flask tightly with a well-fitting cork or rubber stopper. Turn the flask upside down (Fig. 31) and cool it cautiously by pouring cold water on the bottom of it. The water inside begins to boil, and the more you cool it the longer the boiling continues.

The explanation is that by pouring on the cold water you

cool and condense the steam, and so lower the pressure. It is a case of boiling under diminished pressure.

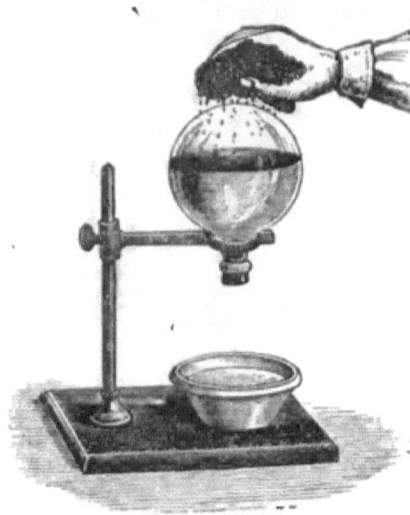

Fig. 31.

The flask used in this and the following experiments should be a round-bottomed flask or 'boltheads'; flat-bottomed flasks readily give way under the pressure. There is no real danger in this, for the flask does not burst outwards, but simply collapses; still it is best to use a small one, holding not more than half a pint (or 300 c.c.)

EXPT. 32.—Connect a flask containing hot water with an air-pump, using an intermediate flask or Woulff's bottle to prevent the steam from getting inside the air-pump. On working the pump the water will be found to boil under the reduced pressure, getting colder as it boils.

A more convenient plan is to use a plain retort instead of a flask, and a water-pump instead of the

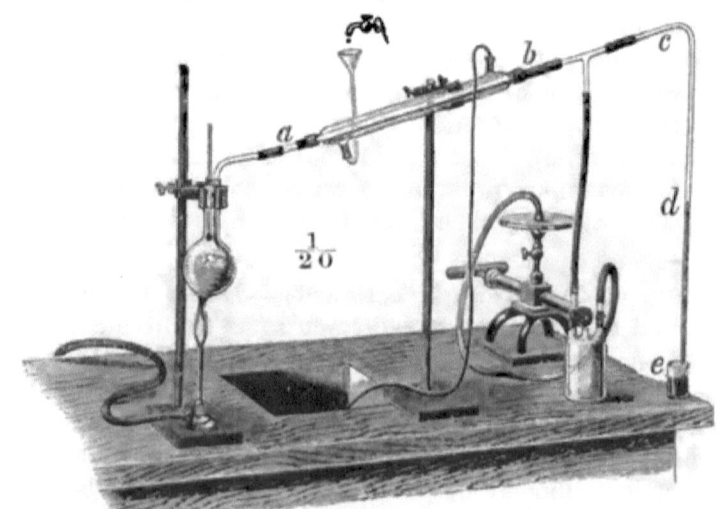

Fig. 32. BOILING UNDER DIMINISHED PRESSURE.

ordinary form of air-pump. The neck of the retort should be drawn out so that it may be pushed into a thick rubber tube to connect it to the water-pump.

EXPT. 33.—In Fig. 32 is shown an arrangement by which water can be made to boil continuously under diminished pressure in a flask connected with a 'reversed condenser' ($ab$). The following are the results of an experiment made with this apparatus—

Temperature at which the water boiled . . . . . 80°
Height to which the mercury was sucked up in the tube $de$  41 cm.
Height of barometer at the time . . . . . 76·5 ,,

The pressure inside the apparatus was therefore 41 cm. less than that outside, or was equal to 76·5 − 41 = 35·5 cm. Thus water boils at 80°, under a pressure of 35·5 cm. We may also express the result by saying that the maximum pressure of water-vapour at 80° is 35·5 cm. By pumping out more air, or by letting a little enter, the pressure can be altered and the boiling-points at different pressures found. It is in this way that the vapour-pressure of water at temperatures above 50° has been measured.

**54. Heat absorbed in Vaporisation.**—When you boil water its temperature does not rise after it has reached the boiling-point, although it is continually heated. What becomes of the heat supplied to the water? It is simply spent in converting the water into steam at the same temperature.

On the other hand, when steam condenses to water, heat is given out. Thus if you hold a cold spoon in a jet of steam issuing from the spout of a kettle, the spoon will rapidly get warm as the steam condenses upon it.

Evaporation at ordinary temperatures, too, is always accompanied by absorption of heat. Illustrations of this cooling effect of evaporation are given in Art. 57.

**55. Latent Heat of Steam.**—It is found that a definite quantity of heat is required to convert a given quantity (say a gramme or a pound) of water at 100° into steam at 100°: this is called the latent heat of vaporisation, and may be defined as follows:—

*The latent heat of vaporisation of water (or the latent heat of steam) is the amount of heat required to convert unit mass (one gramme) of water at 100° into steam at the same temperature.*

The latent heat of steam is 536. By saying this we mean that 536 units of heat are required to convert a gramme of water at 100° into steam without raising its temperature.

The same amount of heat is given out by a gramme of

F

steam in condensing to water at 100°. By observing how much heat is evolved by a known weight of steam in condensing, we can find by experiment the value of the latent heat of steam.

EXPT. 34. *To find the latent heat of steam.*—The method consists in boiling water in a flask and conveying the steam by a bent glass tube into a weighed quantity of water contained in a beaker or calorimeter (such as was used in Expts. 19

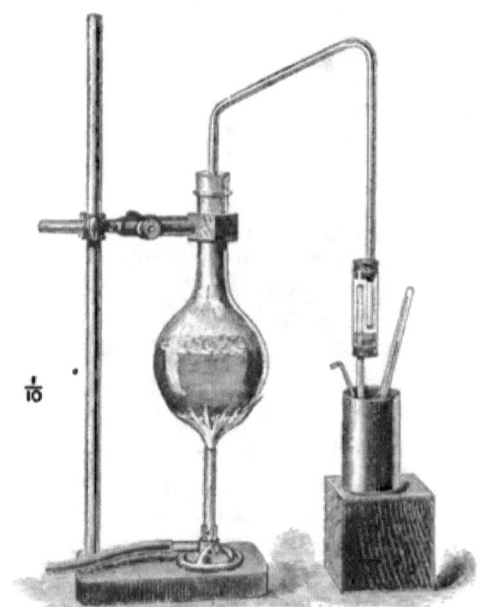

Fig. 33. HEAT EVOLVED IN CONDENSATION.

and 23). The increase in weight of the water tells us how much steam has been condensed, and the rise in temperature tells us how much heat has been given out in the process.

Owing to the cooling action of the air, some of the steam always condenses on its way through the bent tube; the object of the 'trap' shown in Fig. 33 is to catch this condensed water and prevent it from getting into the calorimeter. The trap consists of a wider corked glass tube, through which the steam has to circulate before passing down through the nozzle into the calorimeter; any condensed water remains at the bottom of this wider tube. A vertical screen (of tin-plate or cardboard) should be introduced between the flask and the calorimeter.

Begin by setting the water in the flask to boil; weigh the empty calorimeter; pour into it 100-150 gm. of cold water and again weigh carefully. When the steam is issuing freely from the nozzle take the temperature of the water in the calorimeter. Place the calorimeter in the position shown in Fig. 33, supporting it upon a wood block, so that the nozzle dips well under the water. The steam condenses with a loud rattling noise and the temperature of the water begins to rise. Let this go on until the temperature has risen about 20°, stirring the water throughout the experiment with the thermometer or a wire stirrer. When you want to finish the experiment, take away the block, lower the calorimeter, and remove it from the escaping steam. Do this smartly, and note the highest temperature attained by the water before it begins to cool. Now weigh the calorimeter and water again; the difference between this and the previous weighing tells you the weight of steam condensed. The following are the results of such an experiment—

| | |
|---|---|
| Weight of water in calorimeter . . . | 120 gm. |
| ,, steam condensed . . . . | 5 ,, |
| Temperature of water before passing in steam . . | 10° |
| ,, ,, after ,, ,, . . | 35° |

Now let $x$ be the latent heat of steam: then $5x$ will be the number of heat-units given out by 5 gm. of steam in condensing to water at 100°. The 5 gm. of hot water thus produced will, in cooling from 100° to 35°, give out a further supply of heat amounting to $5 \times 65 = 325$ units.

The effect of this has been to raise 120 gm. of water from 10° to 35°, for which $120 \times 25 = 3000$ heat-units are required. This gives us the equation

$$\underbrace{\text{Heat evolved}}_{\text{in condensing} + \text{in cooling.}} = \begin{array}{l}\text{Heat absorbed} \\ \text{in warming cold water.}\end{array}$$

$$5x + 325 = 3000.$$

Thus $\quad 5x = 3000 - 325 = 2675.$

and $\quad x = 535.$

According to our experiment, then, 1 gm. of steam gives out 535 heat-units in condensing.

**56. Laws of Ebullition.**—We may here sum up the principal facts concerning the ebullition or boiling of liquids.

I. When a liquid is heated it begins to boil at a certain temperature (called its boiling-point), and further heating does not raise the temperature of the liquid, but simply converts it into vapour.

II. This temperature is constant for a given liquid as long as the pressure is constant.

III. When the pressure increases the boiling-point rises, and when the pressure decreases the boiling-point becomes lower.

IV. A definite amount of heat (called the latent heat of vaporisation) is absorbed in converting unit mass of a liquid at the boiling-point into vapour at the same temperature.

**57. Freezing by Evaporation.**— The following experiments illustrate the cooling effect produced by rapid evaporation.

EXPT. 35.—Place a piece of filter-paper on top of the air-thermoscope (Fig. 3, p. 5). Pour some ether on this. The ether rapidly evaporates, and as the heat required for its evaporation is taken from surrounding bodies, it cools the flask and the air inside. The cooling effect is shown by the contraction of the air and the rise of the liquid in the tube.

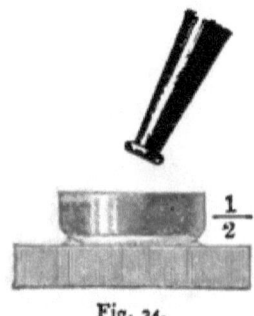

Fig. 34.

EXPT. 36.—Place a watch-glass on a thin piece of wood, with a few drops of water between them; pour ether into the watch-glass, and blow air from a bellows over its surface so as to make it evaporate rapidly. By the time the ether has evaporated the water will have been converted into ice. If you lift up the watch-glass, you will find that the wood sticks to it, the two being frozen together.

The watch-glass may with advantage

be replaced by a shallow capsule, such as may be made of sheet copper (Fig. 34).

EXPT. 37.—Water may be made to evaporate so rapidly as to freeze by its own evaporation. This is most simply done by means of Wollaston's cryophorus (or 'ice-carrier'), a good form of which is shown in Fig. 35. It consists of a glass tube with a bulb at each end. One of these is half-filled with water; the rest of the space contains nothing but water-vapour, the air having been driven out by boiling the water and sealing up the instrument while full of steam.

If the bulb A is surrounded by a freezing mixture (Art. 45), the water-vapour in it condenses; more vapour flows over from B to take its place and is continuously condensed in A. So much heat is taken from B by this process, that in about half an hour its contents will have frozen.

If the water in the cryophorus more than half-fills the bulb B, some of it should be transferred to A: otherwise the expansion in freezing may burst the glass. The freezing mixture should be stirred from time to time. The bulb B should be covered with flannel or with a woollen sock, and it is well to shake it slightly if the freezing does not start in due time.

The cryophorus may be regarded as a distillation-apparatus, of which the freezing mixture and bulb A form the condenser. There is no outside source of heat: the water-vapour in passing over transfers heat from B to A, thus cooling the water in B to the freezing-point, and then converting it into ice.

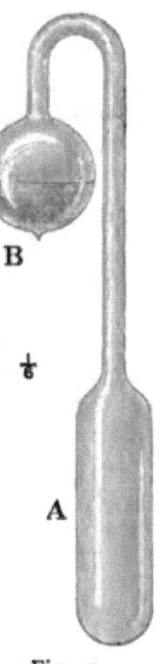

Fig. 35.

**58. Applications and Illustrations.**—Machines for making ice artificially have been constructed on the principles explained above. In Carré's machine a bottle of water is placed in connection with a powerful air-pump and a reservoir containing strong sulphuric acid. After working the pump for a few minutes, the water begins to boil, the vapour is rapidly absorbed by the acid, and the water soon freezes to a solid mass of ice.

The refreshing effect produced by watering a dusty road on a hot day is not due to the laying of the dust alone: the water by its evaporation produces a pleasant coolness. The simplest way of cooling a bottle of wine in hot dry weather is to wrap round it a wet towel; if it is then put to stand in a draught the rapid evaporation of the water will cool it several degrees below the temperature of the air (see Expt. 35).

Our own bodies are cooled by evaporation, which when too rapid produces chills and colds, such as are often caught by exposure to draughts or after taking a warm bath. The air expired from our lungs is always charged with water-vapour, and from the surface of our bodies water is continually removed by perspiration and evaporation. You have probably noticed that a dog when hot after running quickly hangs out his tongue, thus exposing a larger surface for evaporation and increasing the cooling effect.

The lowest temperatures hitherto attained have been produced by the rapid evaporation of liquids with very low boiling-points (liquefied gases).

### Examples on Chapter VIII

1. Explain exactly the nature of boiling. Is it possible to make lukewarm water boil without heating it, and, if so, how?

2. How would you measure the maximum pressure of alcohol-vapour at a temperature of $25°$?

3. How would you distinguish between *vaporisation* and *ebullition*? Does the boiling-point of a liquid depend upon the pressure on its surface? Illustrate your answer with an experiment.

4. How many heat-units are required to convert 50 grammes of water at $12°$ into steam at $100°$?

5. A vessel containing 30 gm. of ice is placed over a spirit-lamp: how much heat will be required to melt it and vaporise the water completely?

6. The calorific power of Welsh steam coal is 8240. How many pounds of water at $100°$ could be converted into steam by the combustion of 1 lb. of this coal?

> The most convenient heat-unit to employ here is the 'pound-degree,' or amount of heat required to raise 1 lb. of water through $1°$. 536 of these units are required to convert 1 lb. of water at $100°$ into steam. In the combustion of 1 lb. of coal 8240 units are evolved, or an amount sufficient to convert into steam $\frac{8240}{536} = 15.4$ lbs. of water at $100°$.

## VAPORISATION AND CONDENSATION

7. 1 lb. of a certain sample of coal is found to be sufficient to evaporate 15 lbs. of water at 100°: how much heat does it give out in burning?

8. How many pounds of steam at 100° will just melt 50 lbs. of ice at 0°?

9. 10 gm. of steam at 100° is condensed in a kilogramme (1000 gm.) of water at 0°, and the temperature of the water is thereby raised to 6.3°: what value does this give for the latent heat of steam?

10. How many grammes of steam at 100° must be passed into 200 gm. of ice-cold water in order to raise it to the boiling-point? What will happen if more steam than this is passed in?

11. Describe an experiment showing that water can be frozen by its own evaporation. What weight of vapour must evaporate in order to freeze a gramme of water at freezing-point?

# CHAPTER IX

## HYGROMETRY

**59. Water-vapour in our Atmosphere.**—The presence of water-vapour in our atmosphere may be shown as follows.

EXPT. 38.—Place some calcium chloride [$CaCl_2$] in a saucer and leave it exposed to the air. The salt becomes wet on the surface and gradually dissolves in the water which it attracts from the air.

Substances like calcium chloride and sulphuric acid, which combine eagerly with water, are said to be *hygroscopic*, and are used for drying air and other gases (see Art. 60).

EXPT. 39.—Dry the outside of a glass tumbler or beaker and pour cold water (or iced water) into it; take it into a warm room. The surface of the glass becomes dimmed. As the air in contact with it is gradually cooled the vapour present is deposited upon it in the form of minute drops of water or *dew*.

You must not conclude from such experiments that the air is fully charged or saturated with water-vapour. This seldom happens; and when it does, the air is incapable of taking up any more water-vapour. Now common observation shows us that, except on very damp days, water exposed to the air rapidly evaporates, as, for example, when wet clothes are hung out to dry. We conclude, then, that our atmosphere generally contains a certain proportion of water-vapour, but seldom sufficient to saturate it. A good 'drying' day is one on which the amount of water-vapour present is far from being sufficient to saturate the air; but the rate of evaporation also depends upon the presence or absence of wind, for when the air in contact with a moist surface is frequently renewed the moisture is more rapidly removed.

The branch of Physics that deals with the moisture of the atmosphere is called *Hygrometry*.

**60. Absolute amount of Water-vapour.**—The amount of water-vapour contained in a given volume of air can be measured by slowly aspirating the air through a series of tubes (Fig. 36) containing a hygroscopic substance which absorbs

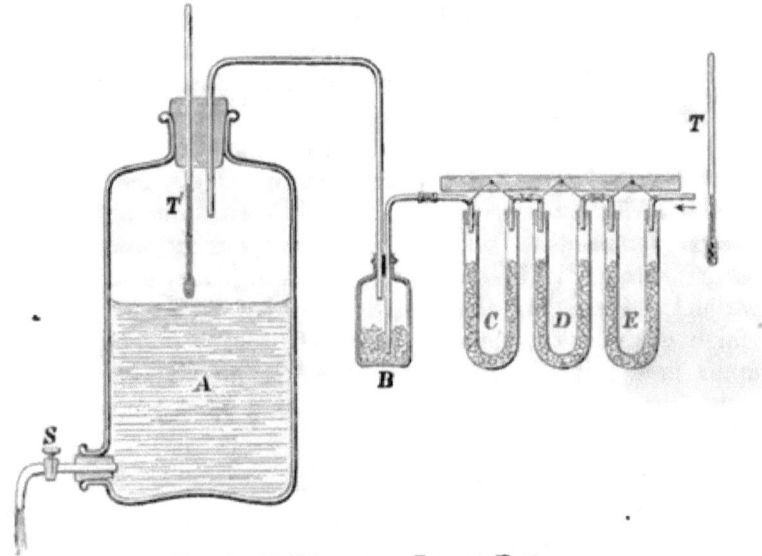

Fig. 36. Aspirator and Drying-Tubes.

and retains the water. Any large bottle or jar with a stop-cock at the bottom will serve as an aspirator (A). This is filled with water which, as it runs out, draws a current of air through the U-tubes C, D, and E. These contain pumice-stone soaked with sulphuric acid, and are carefully weighed before and after the experiment; the increase of weight tells us the total amount of water absorbed. The volume of air which has passed through the apparatus can be found by measuring the water which runs out from the aspirator. The results are expressed by saying that the air contains so many grains of water-vapour per cubic foot, or a certain fraction of a gramme per litre. The quantity thus measured is the absolute (or actual) amount of water-vapour contained in a given volume of the air at the time of the experiment and under the conditions then existing.

Similarly, by passing through the apparatus a known volume of air which has beforehand been saturated with water-vapour, we can find out how much water-vapour is required to saturate a given volume of air at any temperature.

**61. 'Wetness' and 'Dryness.'**—Experiments made in the way described above show that the absolute amount of water-vapour contained in the air at a given time does not fully explain what we call the 'wetness' or 'dryness' of the air. For it is found that on cold misty days in winter, when the air appears quite damp, the amount of water-vapour actually present in a given volume of it is frequently less than on a hot summer day, when we say that the air 'feels dry.' And the reason is not difficult to understand. The amount of water-vapour required to saturate a given space depends upon the temperature, being greater in hot weather than in cold. Now we judge of the 'wetness' or 'dryness' of the air by the rate at which evaporation goes on, and this does not depend chiefly upon how much water-vapour the air already contains, but rather upon *how much more it can take up, i.e.* how far it is from being saturated.

**62. Relative Humidity.**—The apparent wetness or dryness of the air depends, then, upon two things—

(1) upon the amount of water-vapour actually present in a given volume of the air (which we have called the absolute amount of water-vapour or absolute humidity),

and (2) upon the ratio between this and the amount required to saturate the same volume.

This last is called the **Relative Humidity**, and may be defined as follows :—

*The relative humidity of the air at any time is the ratio between the amount of water-vapour actually present at the time and the amount which would be required to saturate it at that temperature.*

Thus, if a given volume of air is found to contain an amount $w$ of water-vapour, and if an amount $W$ would be required in order to saturate it at its actual temperature, then the relative humidity is $\frac{w}{W}$.

This is sometimes expressed as a fraction and sometimes as a percentage; thus, when the air is found to contain one-fifth as much water-vapour as would saturate it, we may either say that the relative humidity is 0·2 or 20 per cent.

**63. The Dew-point.**—If we wish to know the humidity of the air at any time, the simplest way of proceeding is to find out how far it is from being saturated. This can be done by cooling the air gradually down until it begins to deposit its moisture as dew (see Expt. 39). The temperature at which this occurs is called the **dew-point**.

*The dew-point is the temperature at which the air begins to deposit its moisture in the form of dew.*

You should understand clearly that the dew-point is not a *fixed* temperature; it changes continually, as the temperature of the air and the amount of water-vapour contained in it change. On very damp days the dew-point is only slightly below the temperature of the air. In dry weather the air needs to be cooled very considerably before the deposition of dew begins, so that the dew-point is much below the actual temperature.

Instruments used for the purpose of finding the dew-point are called *hygrometers*.

**64. Daniell's Hygrometer.**—This is one of the oldest forms of dew-point instruments. It consists of two bulbs (A and B, Fig. 37) connected by a bent glass tube. The bulb A is half-filled with ether; the rest of the instrument contains nothing but ether-vapour. The bulb B is covered with muslin and can be cooled by dropping ether upon it. This condenses the ether-vapour inside B, and more ether distils over from A to

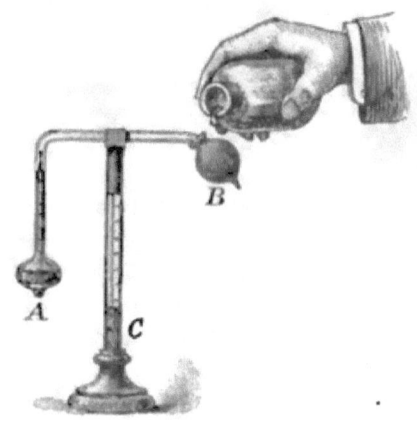

Fig. 37.

take its place. Thus A is gradually cooled, just as the bulb A in the cryophorus (Fig. 34) is cooled when B is placed in a freezing-mixture. At a certain point a thin film of dew makes its appearance on A; the temperature at which this occurs is indicated by a delicate thermometer, the bulb of which dips into the ether in A. Another thermometer attached to the stem (C) gives the actual temperature of the air.

**65. Dines's Hygrometer.**—There are two difficulties in using Daniell's hygrometer. The observer must stand near the instrument; also ether must be used, and this, of course, gets into the air. Both these objections are removed by the hygrometer introduced by Mr. G. Dines (Fig. 38), in

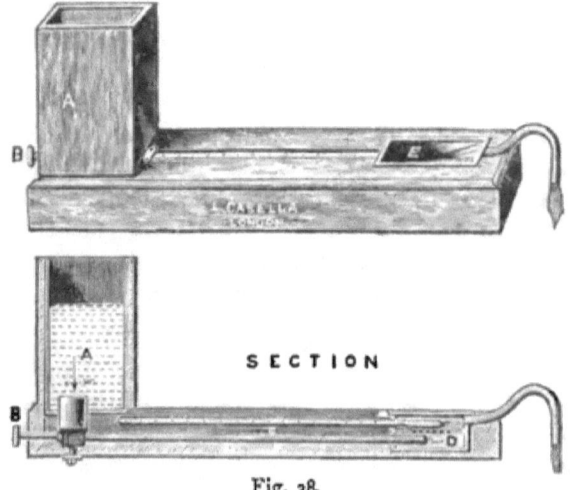

Fig. 38.

which a current of cold water is made to cool a thin plate of black glass (E) upon which the dew is deposited. Cold water (or iced water) is placed in the vessel A, which communicates by a tube with the chamber D. The thin glass plate E forms the top of this chamber, and just underneath it is the bulb of the thermometer C. The observer adjusts the stop-cock B so that a slow current of water flows through; he then watches the plate E, and as soon as its surface becomes dimmed he reads off the temperature indicated by the thermometer C.

It is usual, both with this and the Daniell hygrometer, to make a double observation. First, the temperature at which the deposition of dew begins is observed; then the cooling is stopped, and, as the cold surface gradually gets warmer, the temperature at which the dew disappears is noted. The mean of these two is taken as the dew-point.

**66. Wet and Dry Bulb Hygrometer.**—Another way of judging of the humidity of the atmosphere is by observing the rate at which evaporation is going on. This can be done by measuring the cooling effect produced by it.

EXPT. 40.—Take the temperature of the air with a thermometer, then wrap round the bulb a small piece of fine muslin, and thoroughly wet this with water which has stood for some time in the room and is at the same temperature as the air in it. The thermometer falls, and on a dry day may fall through several degrees. This effect is due to the heat absorbed in evaporation of the water. It is not so marked as if we had used ether (see Expt. 35), but it always occurs excepting when the air is saturated with water-vapour, which is seldom the case.

Upon this principle the wet and dry bulb hygrometer (Fig. 39) is based. It consists of two thermometers, one of which is simply exposed to the air and indicates its temperature, while the bulb of the other is kept continually moist. This is done by covering the bulb with muslin, which is connected by a wick with a vessel containing water. (This vessel should be smaller, and placed at one side; not directly beneath the bulb as in the figure.) The continuous evaporation from the wet bulb

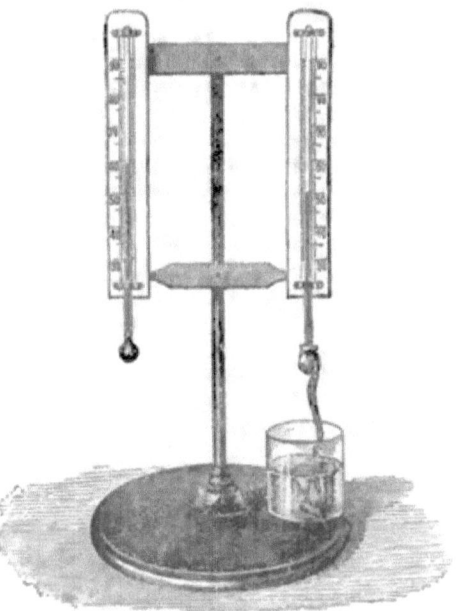

Fig. 39.

keeps its temperature constantly lower than that of the dry bulb. The difference between the two temperatures depends upon the dryness of the air. When this difference is great it indicates that evaporation is going on rapidly, and consequently that the air is dry and the dew-point low. When the difference in temperature is small it indicates that but little evaporation is going on from the wet bulb; hence we conclude that the air already contains much water-vapour and that the dew-point is high.

67. You should understand clearly that the temperature of the wet-bulb thermometer is not constant, but (like the dew-point) changes with the temperature and humidity of the air. Nor must it be confused with the dew-point: it is higher than the dew-point. The greater the difference between the dry and wet bulb readings, the lower the dew-point is.

But consider what would happen if the instrument were surrounded by air already saturated with water-vapour. There would be no further evaporation; the wet-bulb thermometer would show the same temperature as the dry-bulb thermometer, and this (under the conditions stated) would also be the dew-point.

The theory of the instrument has been worked out and a formula has been constructed which enables us to calculate the dew-point when the temperatures of the wet and dry-bulb thermometers are known. The results thus obtained have been compared with those given by the condensation hygrometers (Arts. 64 and 65), and tables have also been drawn up from which the dew-point can at once be found when the readings of the wet and dry-bulb hygrometer are known. On account of its convenience, it is the form of hygrometer most commonly used by meteorologists.

### EXAMPLES ON CHAPTER IX

1. What circumstance determines whether a towel exposed to the air shall dry or shall become damp? Why does a damp cloth exposed to draught become very cold?

2. Describe an instrument for actually determining the dew-point. What do you understand by the term?

# CHAPTER X

## TRANSMISSION OF HEAT—CONDUCTION

**68. Modes of Transmission of Heat.**—Heat may be transmitted from one point to another by three processes, called respectively conduction, convection, and radiation.

You know that if the point of a poker be pushed into a fire, the heating effect is not confined to the point alone: it gradually extends throughout the length of the poker, from the parts near the point towards the handle, which, after a while, may become too hot for you to hold. This transference of heat from hotter to colder parts of the poker, and from the hot poker to your hand, is called **conduction**.

Again, if you withdraw the red-hot poker from the fire and hold your hand above it, you feel a sensation of heat which is mainly due to the warm air (heated by contact with the poker) rising upwards. This process is called **convection**.

Lastly, if you hold your hand an inch or so *below* the red-hot poker you still feel a sensation of heat. This is not due to convection (for heated air always rises upwards), but is produced by a direct transmission of heat through the air from the poker to your hand, a process which is called **radiation**. It is by radiation that your hands are warmed when you hold them in front of the fire.

**69.** *Definition.—Conduction is the transmission of heat from hotter to colder parts of a body, or from a hot body to a colder body in contact with it. This transmission takes place gradually from particle to particle, but without any visible motion of the parts of the body.*

The last sentence is added to point out how conduction differs from radiation and convection. Conduction is a gradual process in which heat only passes from hot to cold parts of the body by heating the intermediate parts. The transmission of heat by radiation is so rapid as to be practically instantaneous; it also takes place without heating the medium through which it passes. In the case of convection heat is conveyed from one point to another by actual motion of the hot body as a whole. These distinctions will be better understood after reading the next two chapters.

**70. Conduction in Solids.**—Solids differ from one another enormously in their power of conducting heat. Metals are generally good conductors, silver and copper being the best. Glass, stone, leather, wood, flannel, and organic substances generally are bad conductors.

EXPT. 41.—Hold one end of a copper or brass wire, about 3 inches long, in a flame. The heat is rapidly conducted along the wire, and it soon becomes too hot to hold. An iron or platinum wire does not get hot so quickly. A strip of wood (a match) can be held until it burns down quite near the fingers. A piece of glass rod or tubing can be held quite comfortably for a long time. It is this bad conducting power of glass that makes glass-blowing possible. When a glass tube is fused in a blowpipe-flame, it can be handled to within an inch or so of the fixed part without discomfort.

EXPT. 42.—Place spoons made of different substances with their bowls dipping into hot water. The handle of a silver spoon (real silver) soon gets so hot that you cannot comfortably hold it. A common spoon (made of Britannia metal) does not get hot so soon; with spoons made of bone or wood the heating effect is scarcely noticeable.

EXPT. 43.—Paper is a bad conductor of heat, and at once becomes scorched and begins to burn when placed in a flame. Yet a piece of paper may be held for some time in a flame without burning (and even without getting scorched) if there is a metal surface immediately behind it to carry away the heat.

This can be well shown by means of a cylinder one half of which is made of brass and the other half of wood (or one end of a wooden cylinder may be turned down so that a brass tube of the same diameter can be slipped over it). Wrap a piece of writing-paper tightly round the cylinder and hold it in a Bunsen flame, as in Fig. 40. The paper covering the wooden half of the cylinder is scorched (just up to the line of junction) long before any effect is produced on the other half. The brass conducts away the heat so rapidly that the paper is kept cool. The experiment shows that brass is a far better conductor of heat than wood is.

Fig. 40.

**71. Comparison of Conducting Powers.**—The following experiments show how the conducting powers (or conductivities) of different materials may be compared.

EXPT. 44.—A number of metal and other rods, of the same length and thickness, are covered with a coating of wax, and are introduced into holes in the front of a metal trough (Fig. 41), which is then filled with boiling water. As the heat travels along each rod and warms it up to the proper temperature, the wax melts.

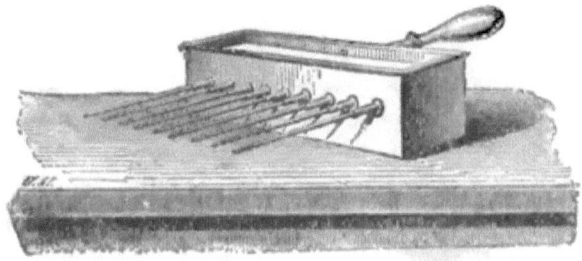

Fig. 41.

Wait until there is no further sign of melting, and then observe that the wax has melted much farther along some rods than others.

EXPT. 45.—Take two strips or bars of iron and copper respectively, and of the same size. Attach small wooden balls (or marbles) to them by means of wax. Fix them end to end

Fig. 42.

and place a burner underneath, so as to heat them equally (a better plan is to rivet both to a cross-piece and heat this).

The balls will be found to drop off the copper strip more rapidly than off the iron.

EXPT. 46.—Cut discs of copper, iron, wood, and cork of about $\frac{1}{4}$ inch thickness, and large enough to cover the top of the air-thermoscope (p. 5). As a source of heat use a metal cylinder (iron or copper) heated to 100° by immersion in boiling water.

Place one of the discs on top of the air-thermoscope, and on this put the hot cylinder. Wait a couple of minutes and observe the greatest depression produced. Try the other discs in the same way. Copper gives the greatest depression; wood and cork scarcely allow any heat to pass.

72. When one end of a bar is heated, the rate at which its temperature rises depends not only upon its conductivity but also upon its specific heat. Other things being equal, a bar having a low specific heat would get hot more rapidly than another having a high specific heat. Thus the above methods of experimenting are not quite satisfactory, excepting where the substances compared have nearly the same specific heat (as in the case of iron and copper, Expt. 45). But they show sufficiently well that metals are, in general, good conductors of heat, silver and copper being the best; whereas stone, glass, wood, cork, etc., do not conduct heat nearly so well, or are bad conductors.

In order to keep our bodies warm we make use of such bad conductors (woollen clothes, flannel blankets, fur, leather, etc.) Other applications of non-conductors will readily occur to you —*e.g.* kettle-holders, the handles of coffee-pots, kettles, and soldering-irons, etc. The badly-conducting materials which we use for 'keeping out cold' are also the best for keeping out heat. Thus, if you wish to keep a block of ice in hot weather, and have no ice-chest to put it in, it should be wrapped up in flannel.

You will now be better able to understand why it is that, in our climate, metals and other good conductors feel cold to the touch (see pp. 1-3). If the temperature of surrounding objects were higher than that of our bodies, the reverse would be the case; and so in the hot rooms of Turkish baths it is found that iron and stone are painful to touch because they part with heat so readily; whereas wood and flannel can be handled without discomfort.

**73. Action of Wire-gauze on Flame.**—A combustible substance will not burn, even in presence of air or oxygen, unless it is raised to a certain temperature, which is called the 'temperature of ignition' of the particular substance. Now when a good conductor is introduced into a flame it rapidly withdraws heat from the flame and thus cools it. The following experiments show that it is easy to cool and extinguish a flame in this way.

EXPT. 47.—Coil some copper wire round a rod, leaving a piece free at the end to serve as a handle, and making the coil of such size as to fit round the flame of a spirit-lamp. Lower the coil vertically over the flame until the bottom of it touches the wick. The flame shrinks away from the coil and then goes out.

The coil does not crush out the flame; there is plenty of room for it to burn and air can easily pass between the turns of wire. That the effect is really due to the cooling action of the metal may be shown as follows. Heat the coil to redness by means of a Bunsen burner and lower it again over the flame of the spirit-lamp; it is no longer extinguished. Cool the coil by dipping it in water; dry it, and repeat the experiment. The cold coil at once puts out the flame.

[A coil of No. 16 wire, $\frac{5}{8}$ in. in diameter and 2 in. long, does well for a spirit-lamp. For a candle-flame use a smaller coil.]

EXPT. 48.—Procure a piece of wire-gauze, such as is commonly used in laboratories for supporting glass vessels in which water is to be boiled. Lower the gauze upon the flame of a Bunsen burner. The flame does not pass through the meshes of the gauze but appears to be crushed down by it (Fig. 43).

Fig. 43.

Fig. 44.

The gauze cools the mixture of coal-gas and air that streams through it below the temperature of ignition. That this inflammable mixture does readily pass through the gauze may be shown by setting fire to it with a lighted match. It may also take fire of itself if the gauze gets red hot.

EXPT. 49.—Turn on the burner but do not light the gas. Place the wire-gauze on the top of the burner and light the gas above it. The gauze may now be lifted an inch or so above the burner (Fig. 44), but the flame does not strike down.

**74. The Safety-lamp.**—Upon this principle depends the action of the safety-lamp invented by Sir Humphry Davy. It consists of an oil-lamp (Figs. 45, 46), the flame of which is surrounded by a cylinder of wire-gauze closed on top by a brass plate. The gauze does not interfere with the burning of the oil, but, as we have seen, it prevents any flame from passing from the inside to the outside. Thus the lamp can be safely used even in an atmosphere containing inflammable gases, such as the 'fire-damp' which is the terror of coal-miners.

The use of naked lights is no longer permitted in mines which are known to be 'fiery' or liable to accumulations of fire-damp. By using the Davy lamp the miner is able to work with safety in such mines, and is also provided with a warning of the approach of danger. For when the amount of inflammable gas present becomes great enough to produce an explosion, it burns inside the gauze with a blue flame, and the miner then knows that it is time for him to withdraw.

EXPT. 50.—Connect a glass tube by india-rubber tubing to a gas tap, and direct the stream of gas against the gauze of a lighted safety-lamp. The flame of the lamp grows larger, and the coal-gas burns quietly inside, but the flame does not spread outside the gauze.

Fig. 45.

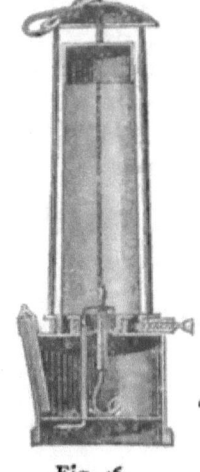

Fig. 46.

**75. Conduction in Liquids.**—As may be seen from the following experiments, liquids are bad conductors of heat. Mercury and other molten metals are exceptions to this rule.

EXPT. 51.—Apparatus required—A differential thermometer of the form shown in Fig. 47 (having one bulb higher than the other): a vessel large enough to contain the thermometer; and a tin dish which can be supported from the top of this.

Pour into the vessel enough water to cover the bulb to a depth of a couple of inches. Support the tin dish so that it just dips into the water, and

Fig. 47.

then fill it with boiling water or hot oil. It will be some time before enough heat is conducted downward to affect the thermometer.

EXPT. 52.—Another form of the experiment is shown in Fig. 48. An ordinary (mercurial) thermometer is inserted through a hole in the side of a vessel into which water is poured to about an inch above the bulb. The water is heated by carefully pouring hot oil on its surface, or a few drops of

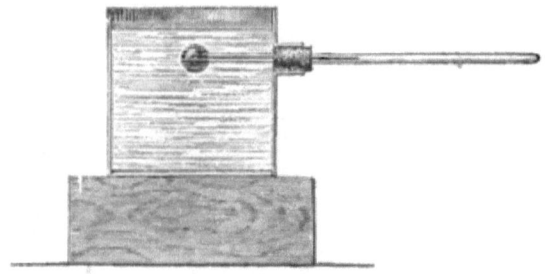

Fig. 48.

benzene may be poured on its surface and ignited; in this case the thermometer will scarcely be affected at all.

EXPT. 53.—Twist some copper wire round a piece of ice, so as to make it heavy enough to sink in water; drop it into a long test-tube nearly filled with water; tilt the test-tube so that the water in the upper part may be heated over a flame. The water may thus be made to boil on top without causing any noticeable melting of the ice.

**76. Mercury a good Conductor.**—That mercury conducts heat much better than water may be shown by the following experiment.

EXPT. 54.—Take two test-tubes of equal size and fix a wooden ball (or a marble) on to the bottom of each with beeswax. Nearly fill one with water and the other with mercury.

Bend a piece of thick copper wire to the shape shown in Fig. 49 (which is drawn one-fifth of the actual size). Heat the wire to redness, and support it with one leg dipping into the water and the other into the mercury. The ball will soon drop off the left-hand tube, while the other will remain on although the water may be spluttering and boiling on top.

Observe that in all the above experiments the liquid is heated from on top, so that the heat has to pass downward; this is the only way in which the true conducting power of liquids can be examined. If they are heated from below, convection-currents are produced; and the same is true for gases (see next chapter).

**77. Conduction in Gases.**—Gases are even worse conductors of heat than liquids. Hence ice-houses (for the storage of ice in summer) are constructed with double walls: the air between these protects the contents from external heat. So in cold climates double doors and windows are used to retain heat and keep dwelling-houses warm.

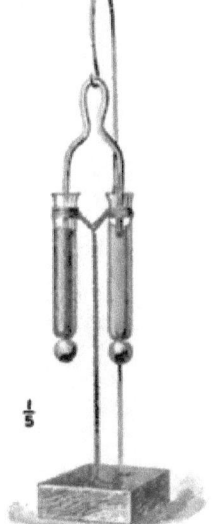

$\frac{1}{5}$

Fig. 49.

EXPT. 55.—Place upon the palm of your hand a *loose* pile of some badly-conducting powder such as lime, asbestos, or plaster of Paris. On this lay the point of a red-hot poker or a piece of iron (not too heavy) heated to redness. The iron can be held for some time without burning the hand.

If the powder were pressed down *tightly* upon the hand (as would happen if the iron were heavy), it would soon feel painfully hot. This shows that the effect is mainly due to the bad conducting power of the air entangled between the solid particles. The powder is interposed to cut off the direct radiation of heat.

The low conducting power of air is made use of in the construction of ice-safes, which are made with a double casing. The space between is filled with straw, sawdust, or other loose, badly-conducting material, which hinders the free motion of the air, or in other words, stops convection. Fur, feathers, eider-down, and wool owe their warmth to the air entangled in them.

# CHAPTER XI

## TRANSMISSION OF HEAT—CONVECTION

**78.** On account of their low conductivity, fluids (*i.e.* liquids and gases) are only warmed very slowly when the heat is applied from on top. But they soon get warm when the heat is applied from below, as when water is heated in a kettle by placing it on a fire, or when air is heated by contact with warmer soil. The parts which are first heated expand, and being thus rendered lighter [1] they ascend: their place is taken by colder parts of the fluid, which, in turn, are heated and also ascend. There is thus produced in the fluid an upward current which carries heat with it. This mode of transmission of heat is called convection, and may be defined as follows:—

*Convection is the transmission of heat by actual motion of the parts of a heated fluid.*

It is clear that convection-currents can only be produced in liquids and gases, for the parts of a solid cannot move about in the manner described.

**79. Convection in Liquids.**—The convection of heat in liquids may be illustrated by the following experiments.

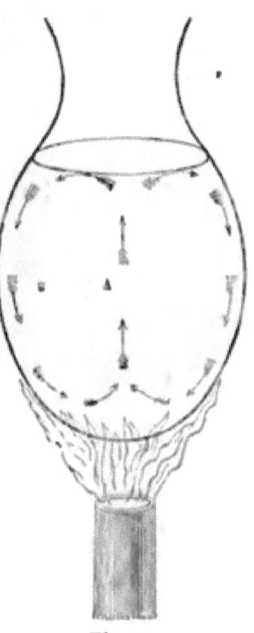

Fig. 50.

EXPT. 56.—Fill a round-bottomed flask with water, and

[1] See Expt. 12, p. 25.

drop into it some crystals of magenta dye. Heat the flask over a small flame (smaller than that shown in the figure). The water just above the heated spot ascends, and its place is taken by colder water from the sides. A current of hot (and coloured) water rises up the middle of the flask, spreads out on top, and then works its way down the colder sides as indicated by the arrows.

The direction of the currents may also be shown by throwing in sawdust.

EXPT. 57.—With the aid of a lantern convection-currents in water may be shown by projection on a screen.

Fig. 51.

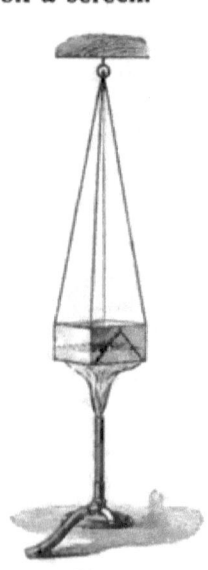

Fig. 52.

The water (slightly warmed) is placed in a cell with parallel glass sides (Fig. 51), and the image of this is focussed on the screen. If a piece of ice be now floated on the water, it will gradually melt, and the colder water sinks to the bottom of the cell. The descending current shows upon the screen in the form of streaks, which are due to the unequal refraction of cold and warm water (see Light, ch. vi.)

Or again, the cell may be filled with cold water and hot water may be introduced into this by means of a pipette dipping into it. If this is done very slowly and cautiously, the hot water will be seen to rise upwards and form a layer floating on the colder water. (The image on the screen is of course inverted.) (See Light, ch. viii.)

EXPT. 58.—Make a paper box by folding stout writing-paper; put a few stitches in to keep the flaps from opening, and hang the box up by threads from the four corners (Fig. 52). Half-fill the box with water and let the flame of a Bunsen burner play on the bottom of it. The water can be heated to boiling without scorching the paper.

You should not conclude from this experiment that paper is a good conductor of heat. It rather shows how rapidly heat is carried off by convection. If it were not for the water the paper would soon burn, and if the flame is too big the box does get scorched above the water-line.

**80. Circulation of Water.**—The following experiment shows how a continuous circulation of water can be maintained by convection.

EXPT. 59.—A glass tube of the form shown in Fig. 53 is

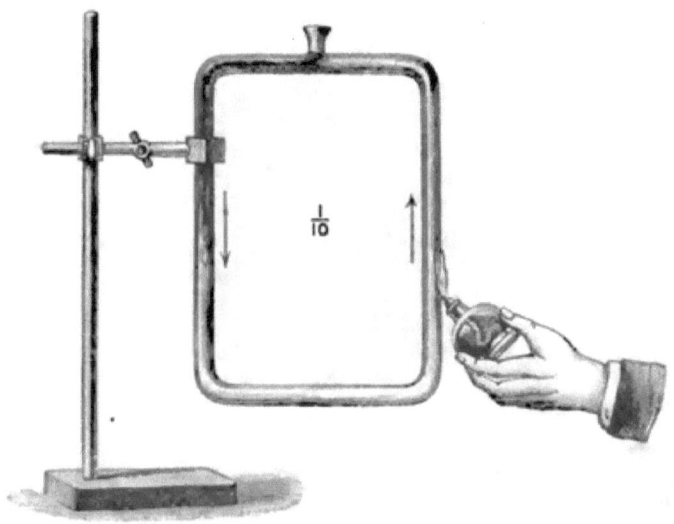

Fig. 53.

filled with water up to the open neck on top, and one of the vertical branches of the tube is heated (near the lower corner) by means of a small gas-flame or spirit-lamp. The heated water ascends and its place is taken by colder water from below.

To show up the motion of the water better, drop in through the neck a little magenta dye (as much as can be taken up on

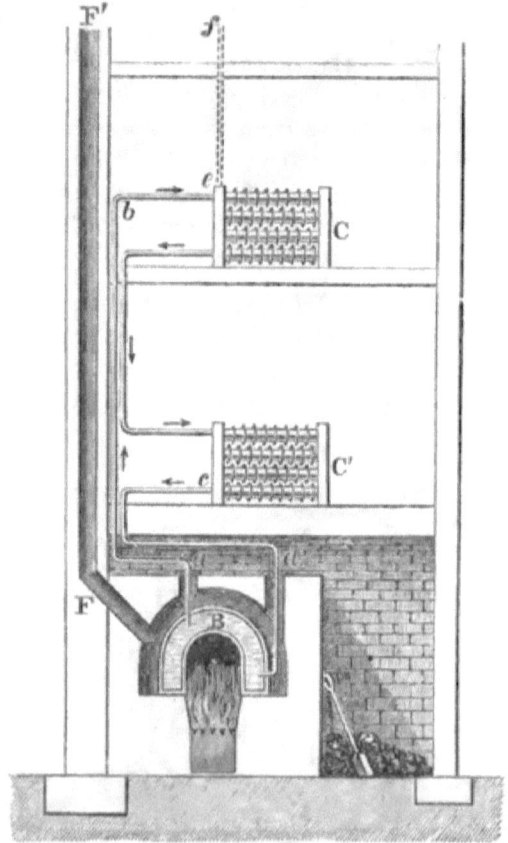

Fig. 54. Hot-Water Heating Apparatus.

the point of a penknife, crushed and moistened with water before dropping it in). The coloured water moves towards the left, passes down the left-hand tube, and so round as shown by the arrows. The circulation can be kept up as long as you please, but cannot well be followed after the dye has worked round to the neck.

**81. Heating by Hot Water.**—One of the best methods of heating large buildings is by means of hot-water pipes. The general arrangement of a hot-water system is shown in Fig. 54. The principle on which the method depends has been illustrated in Expt. 59.

The water is heated in the boiler B, and rises up through the pipe $ab$; it then passes through coils (C, C', etc.) which are placed in the various rooms and serve to distribute the heat wherever it is most required. Having parted with much of its heat, the water is now colder and heavier; it sinks through the return pipe $cd$ to the bottom of the boiler, where it is again heated and begins its journey afresh.

All three processes—conduction, convection, and radiation—come into play here. Heat is radiated out into each room from the pipes and coils; these are painted a dull black so as to increase their emissive power (p. 103), and the separate pipes in each coil are provided with numerous flanges, so as to increase the radiating surface. It is by conduction that the heat passes from the furnace through the boiler into the water and, later on, from the water to the outer surface of the pipes and coils. The air above these is heated by convection; and the whole system is a capital example of a continuous water convection-current carrying heat with it from the boiler to the coils.

**82. Convection in Gases.**—The draught up a chimney is a familiar example of convection. Before passing into the chimney the air is heated by the fire and is thus rendered lighter than the air outside. In the same way convection-currents are produced in the chimneys of oil-lamps.

The upward motion of the air above any hot body (*e.g.* a red-hot poker) may be observed by holding near it a piece of smouldering brown paper (or 'touch-paper' made by soaking brown paper in saltpetre solution and then drying it).

The quivering appearance of objects seen over lime-kilns and chimney-stacks is due to the irregular refraction of light through the heated air rising up. In this way convection-currents in air may be well shown by using a lantern as in Expt. 57 (excepting that here no focussing is necessary).

EXPT. 60.—Hold a spirit-lamp in a strong beam of light from the lantern, and examine its shadow on the screen. Wavy streaks of light around and above the flame show the ascending currents of hot air.

Examine in the same way the shadow of a red-hot poker or

of an iron or copper ball heated to redness. Observe that *below* the ball the air is only heated to a very short distance. This shows what a bad conductor air is.

EXPT. 61.—Place a lighted candle-end on a saucer and pour some water round it; over the candle place a lamp-glass. The flame flickers awhile and then goes out. There is an open outlet above for the heated air and products of combustion, but no fresh air can get in below and so the candle cannot burn.

Repeat the experiment and introduce a piece of cardboard down the middle of the lamp-glass as in Fig. 55. The flame appears to be blown about a bit at first, but it brightens up and keeps alight. The cardboard has roughly divided up the chimney into two halves and the flame makes use of one of these to get rid of the hot gases while the other brings fresh air down to it. The existence of these two currents can be shown by holding smouldering paper near the top of the lamp-glass.

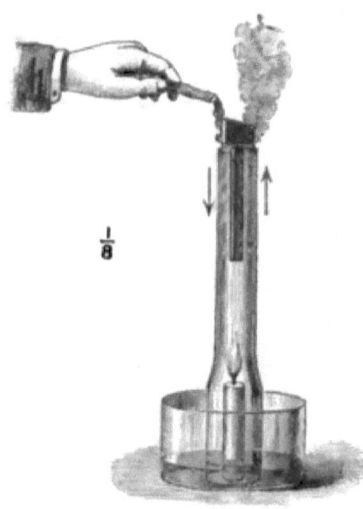

Fig. 55.

83. **Ventilation of Rooms.**—In our climate dwelling-rooms are usually warmer than the air outside. This is not only due to fires and lights, but also to the heat given out by our own bodies. This hot and impure air rises to the top of the room, as you may easily prove by standing on a table in a closed room where several gas-jets have been burning. In order to keep the air sufficiently pure for respiration we require some system of ventilation, *i.e.* some means of getting rid of hot impure air and replacing it by cool fresh air.

If you slightly open the door of a room and hold a lighted candle in the gap, you will generally find that near the floor the flame is blown inwards (Fig. 56). At the top (unless the room is very high) it is drawn outwards: while a position (about half-way up) can generally be found in which the flame burns steadily. Thus cold air tends to make its way into the room at the bottom, driving out the warm light air above. Two things are therefore necessary before a room can be properly ventilated: an outlet for warm air just under the ceiling, and an inlet for fresh air near the floor.

**84. Ventilation of Mines.**—Coal-mines have to be thoroughly ventilated in order to supply fresh air for breathing and also to prevent accumulation of the dangerous and inflammable fire-damp referred to in Art. 74. For this purpose two vertical shafts are provided at opposite ends of the mine. At the bottom of one of these (called the up-cast shaft) a large fire is kept burning, and this creates a powerful upward draught. Fresh cold air enters through the other shaft (called the down-cast shaft), and has to make its way through the various workings of the mine before it reaches the up-cast shaft. The action of the system may be illustrated as follows:—

Fig. 56.

EXPT. 62.—Take off the front of a flat wooden box and replace it by a sheet of glass. Cut two holes in the top of the box (at opposite ends) and put a lighted candle in the box under one of these. Over each of the holes place a lamp-chimney or wide glass tube, as in Fig. 57. The left-hand tube represents the up-cast shaft, the candle the fire, the right-hand tube the down-cast shaft, and the box itself the mine.

By holding smouldering paper at the top of the tubes, it can be shown that there is a steady current of air down the right-hand tube, through the box, and up the left-hand tube.

**85. Convection-Currents in Nature.**—The student should here read Arts. 24, 25 again. Winds, which are natural convection-currents on a large scale, will be treated of in Chapter XIII, after we have considered radiation.

Ocean currents may be regarded as convection-currents produced by the unequal heating of the surface of our globe.

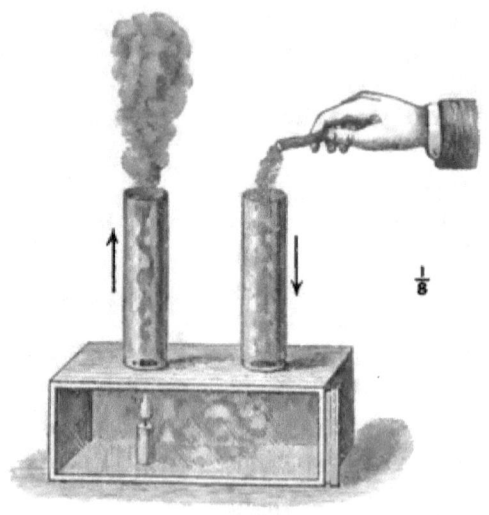

Fig. 57.

In general they follow the direction of the prevailing winds. These in tropical regions are easterly (*i.e.* blow from east to west), and so in the Atlantic Ocean there is an equatorial current from east to west. This passes along the north-east shoulder of South America, into the Caribbean Sea, and flows out of the Gulf of Mexico as a mighty stream of warm water—the **Gulf Stream**—which skirts the coast of the United States and then sweeps across the Atlantic to the north-west coasts of Europe. As an example of the effect of the Gulf Stream on climate, it may be stated that the harbour of Hammerfest in Norway is free from ice all the year round; whereas the mouth of the Baltic (12 degrees farther south) and the river Hudson (in the same latitude as Rome) are frozen over three months in the year. The comparative mildness of winter in the British Isles is largely due to the influence of the Gulf Stream.

# CHAPTER XII

## TRANSMISSION OF HEAT—RADIATION

86. We have already referred (Art. 68) to radiation as a third mode of transmission of heat. It is by radiation that the sun warms our earth. We may at once mention two respects in which radiation differs from the other modes of transmission of heat. A hot body emits radiation *in all directions* and *in straight lines*. An obstacle interposed in the direct line between you and a source of heat at once cuts off the radiation. You can protect your face from the heat of a fire by holding a book or paper between; and when the sun's rays are too powerful you seek relief by getting into the shade.

The transmission of heat by convection, on the other hand, always takes place in one direction (*e.g.* by upward currents); and conduction is not restricted to straight lines, for a bent wire conducts heat as well as a straight one.

Light is known to travel at the rate of about 186,400 miles per second. Now, when an eclipse of the sun takes place, it is found that the light and heat are cut off at the same time; hence both must travel with the same enormous speed. Another characteristic of heat-radiation is that the medium or substance through which it passes is not thereby sensibly warmed. It is true that many media are only partially transparent to heat; these absorb a portion of the radiation as it passes through them (just as partially transparent glass absorbs a certain proportion of light), and are thereby warmed. With this reservation we may define the process of radiation as follows :—

*Heat is said to be transmitted by radiation when it passes from one point to another in straight lines, with great speed, and without heating the medium through which it passes.*

**87** Much of what is stated in this chapter will be better understood after the student has read the next section of this book (Light). The laws of reflection and refraction are the same for heat as for light. If an image of the sun be formed with a lens on paper or on the back of the hand, the heat is found to be focussed at about the same spot, a fact with which schoolboys are sufficiently familiar, and which shows that heat-radiation can be refracted as light is. Again, the law of inverse squares (p.124) applies here as in the case of light, and it is best to study the two sets of phenomena in connection with one another.

A body heated in a dark room to a comparatively low temperature, say below a dull-red heat, is not visible; it emits only heat-radiation. But when it is heated above this temperature it becomes visible; it now emits both kinds of radiation, or it would be more correct to say that the effect produced by its radiation depends upon the object on which that radiation falls. It affects the surface of our bodies as heat; it produces in our eyes the sensation of light.

The common use of the term 'radiant heat' might lead the student to suppose that there are different kinds of heat, of which 'radiant heat' is one. This is not so. Strictly speaking, radiation is not heat at all: at any rate, it does not exhibit the ordinary properties of heat. If it were, it could not pass, as it is known to do, through bodies without heating them. There is reason for believing that radiation travels in the form of waves, much as waves are propagated on the surface of water by the upward and downward motion of the particles of water. The disturbing cause producing these waves is the hot body emitting the radiation. The waves, in general, travel freely outwards in all directions; but when they strike against an obstacle they may be partly reflected from its surface, partly absorbed by it, and partly transmitted through it. Bright polished tin-plate may be taken as a type of a good reflector; lamp-black as a type of a good absorber; and rock-salt as the best example of a substance which is transparent to heat-radiation.

**88. Apparatus, etc.**—For detecting the presence of heat-radiation the differential thermometer (Fig. 19) may be employed. The bulbs should be painted over a dull black;[1] the reason for this will be explained in Art. 90. The ether-thermoscope shown in Fig. 58 is even more delicate than the differential thermometer. It is made like the cryophorus (p. 69), excepting that it contains coloured ether instead of water. It may be regarded as a differential thermometer in which the ether forms the indicating column and in which the bulbs contain ether-vapour instead of air. When the lower bulb is held in the hand or placed near a hot body the pressure of the ether-vapour inside it increases (p. 59), and so the liquid ether is forced up in the stem. The lower bulb should be blackened.

As a source of heat a cubical tin box (of about 5 inches side, see Fig. 64) filled with water may be used. One of the vertical sides of the box should be left as bright tin-plate; the opposite one may be blackened as explained in the footnote or by holding it over the smoky flame of burning camphor or turpentine. The other two may be treated as required for any particular experiment; *e.g.* you may paste white paper over one and roughen the other with coarse sandpaper. A box so prepared is called a 'Leslie's cube,' for such boxes were used by Sir. J. Leslie in his researches on radiation.

Fig. 58.

As a more powerful source of heat we may use a copper or iron ball (about $1\frac{1}{2}$ inch in diameter) heated to redness in a fire (Fig. 62).

We will now proceed to consider the reflection, absorption, and emission of radiation, in so far as they can be illustrated with the aid of simple apparatus.

**89. Reflection.**—EXPT. 63.—Procure two tubes of bright tin-plate about 2 feet 6 inches long and 3 inches in diameter. Support them horizontally on suitable stands, as shown in Fig. 59, the two tubes making an angle of about 120° with one another. At the mouth of one place the ether-

---

[1] A lamp-black surface is best. The lamp-black should be ground up with thin shellac varnish, or with a mixture of equal parts of gold size and turpentine, and then applied with a brush.

thermoscope T (or one bulb of the differential thermometer), and at the mouth of the other place the red-hot ball.

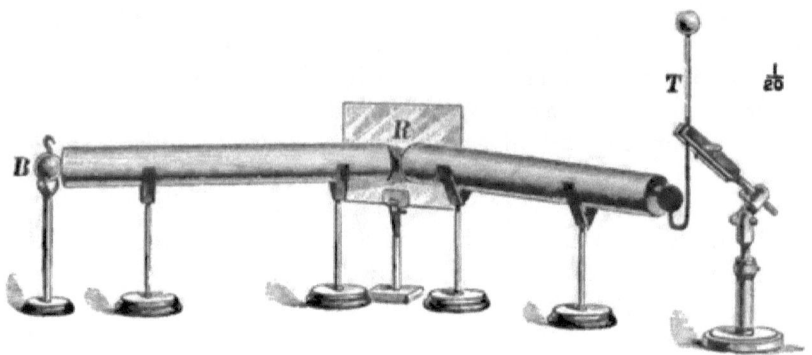

Fig. 59.

Generally, no effect will be produced on the thermoscope. Now place at R a vertical sheet of bright tin-plate equally inclined to the two tubes. This reflects the radiation through the second tube, and the heating effect is shown by the rapid rise of the liquid column in the thermoscope. By altering the position of the tin-plate sheet, it can be shown that the heating effect is greatest when the reflector is equally inclined to the two tubes.

If NP (Fig. 60) be the normal or perpendicular to the reflecting surface where the incident ray IN meets it, the angle INP is called the angle of incidence. The angle PNR, between the reflected ray NR and the normal, is called the angle of reflection. The laws of reflection here are the same as in the case of light (p. 137), viz.—

I. The reflected ray lies in the plane containing the incident ray and the normal, and on the opposite side of the normal.

II. The angles of incidence and reflection are equal.

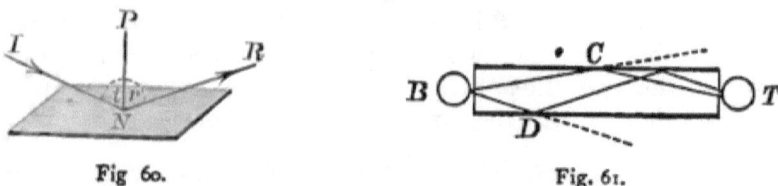

Fig 60.  Fig. 61.

EXPT. 64.—The use of the tin-plate tube is in itself an example of reflection. If the ball and thermoscope be placed as at B and T (Fig. 61), with the tube between them, the heating effect will be so great as to drive the ether into the upper bulb. On removing the tube the heating effect becomes much smaller. Only a small fraction of the total radiation (viz. that travelling directly from B to T) now reaches the bulb. But when the tube is interposed, rays such as BC and BD (which would otherwise travel along the dotted lines) strike against it, and after one or more reflections from the inside of the tube they reach the bulb T.

EXPT. 65.—A more striking illustration of reflection can be obtained with a pair of concave metallic mirrors (best of nickel-plated copper). These should be placed facing one another in a straight line. The red-hot ball is placed in the principal focus of one and the bulb of the thermoscope in the focus of the other (Light, ch. v.) Even when the mirrors are over 10 feet apart the reflection and concentration of the heat can in this way be easily shown.

**90. Absorptive Power.**—A sheet of bright tin-plate reflects most of the radiation which falls upon it. A sheet which has not been tinned ('black-plate'), or which has had its surface blackened, is a very poor reflector: it *absorbs* most of the radiation and is thereby heated. Thus different substances have different absorptive powers. The following experiments show that dull black surfaces are good absorbers, whereas bright polished surfaces are bad absorbers.

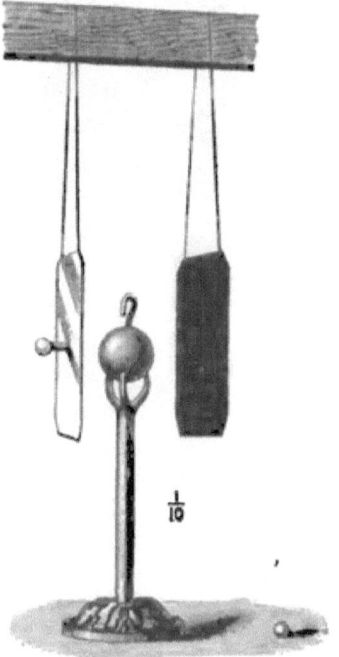

EXPT. 66.—Take two sheets of tin-plate and blacken the face of one with lamp-black (p. 99). On the back of each (in the centre) solder a short, thick copper rod. To the end of each rod attach a marble or wooden ball with wax. Hang up the sheets as shown in Fig. 62, and place the red-hot ball midway between them. The ball attached to the blackened sheet drops off long before the other does. This proves that the lamp-black surface is a much better absorber than the bright tin-plate.

Fig. 62.

EXPT. 67.—Blacken one bulb of the differential air-thermometer and leave the other clean. Expose them to the same source of heat and at equal distances. The blackened bulb becomes hotter than the other, proving that lamp-black is a better absorber than glass.

EXPT. 68.—Paint a black ring (Fig. 63) with lamp-black varnish on the face of a stout piece of tin-foil. Grind up some mercuric iodide with weak gum-water (or mastic varnish), and paint this over the back of the

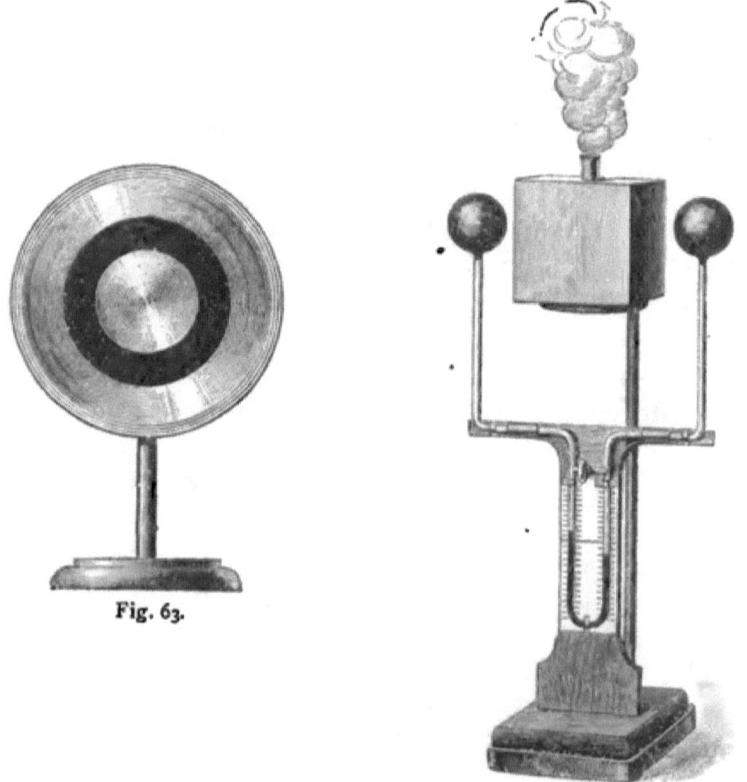

Fig. 63.

Fig. 64.

tin-foil. Mercuric iodide is a scarlet crystalline powder; but when heated to about 150° it is converted into a yellow modification.

Now heat a flat piece of iron to redness and hold it about 2 inches in front of the tin-foil; on looking at the back of it you will see on the scarlet background a ring of *yellow* mercuric iodide corresponding exactly to the blackened ring in front.

**91. Emissive or Radiating Power.**—The radiation emitted by a hot body depends (1) upon its temperature, and (2) upon the nature of its surface. The radiation increases as the body gets hotter, but it does not depend upon the temperature alone. Some substances give out more radiation than others at the same temperature, and are said to have higher emissive or radiating powers. And it is found that the surfaces which have the highest emissive powers are precisely those which are the best absorbers (*e.g.* dull black surfaces). This is expressed by saying that '*good absorbers are also good radiators.*' It will be sufficient if we test the correctness of this statement in the case of the bright and black tin-plate.

EXPT. 69.—Turn up the blackened bulbs of the differential thermometer as shown in Fig. 64. Place the Leslie's cube on the ring of a retort-stand mid-way between the bulbs. Turn it so that the lamp-blacked surface faces one of the bulbs and the bright tin-plate surface faces the other. See that the bulbs are at equal distances from the sides. Fill the cube with boiling water. The radiating surfaces are now at the same temperature (100°); but the liquid column on the same side as the blackened surface at once begins to fall. This shows that lamp-black is a much more powerful radiator than bright tin-plate.

EXPT. 70.—Procure two tin pots (saucepans) of the same size, and coat the outside of one with lamp-black (p. 99), leaving the other bright; hang them up by strings at a convenient height and distance apart, so that the bulbs of the differential thermometer can be introduced, one into each pot. (For this experiment the bulbs should be turned down as in Fig. 19, and they need not be blackened.) Open the stop-cock of the thermometer; take enough water to fill both pots and warm it to about 70°; pour half into one pot and half into the other. Close the stop-cock and watch the thermometer as the water cools. The column on the same side as the blackened pot gradually rises, showing that this cools more rapidly than the bright one. Thus we have again proved that a lamp-black surface radiates out more heat than a surface of bright tin-plate at the same temperature.

Just as the surface of the earth is a better absorber than the surface of the sea, so it is a better radiator. Both, during the night, lose heat by radiation into space. But the earth cools more rapidly than the sea does: and, of the various bodies on the surface of the earth, those which are the best radiators will suffer the greatest loss of heat. This is one reason why dew is found to be more copiously deposited upon grass and plants than upon stones and metal surfaces (Art. 95).

**92. Diathermancy or Transmissive Power.**—Bodies which easily allow heat radiation to pass through them are said to be diathermanous. Among solids rock-salt is the best example of such a substance. Our atmosphere is fairly diathermanous; but water-vapour is not so. The presence of moisture in the atmosphere hinders the radiation of heat, and hence the loss of heat by radiation during the night is more marked in dry climates than in places where the air is moist.

Although diathermancy to heat-radiation corresponds to transparency to light, it by no means follows that bodies transparent to light are also diathermanous or transparent to heat-radiation. Thus alum is transparent to light, but is opaque to heat-radiation. The same is true for water; so that a cell containing water may be used for cutting off heat while it lets the light through.

If iodine be dissolved in carbon bisulphide, a reddish-black solution is obtained which (except in thin layers) is opaque to light. Yet this solution readily transmits heat-radiation, and a flask filled with it may be used for concentrating heat just as a glass lens or a flask filled with water may be used for focussing light.

The behaviour of glass is peculiar. It stops the radiation from a red-hot ball or from a fire; a fact which is sufficiently illustrated by the use of glass fire-screens. But if the source of heat be at a higher temperature (*e.g.* the sun), a considerable fraction of the radiation from it is transmitted through a sheet of glass. If you put your hand on the inner ledge of a window through which the sun is shining, you may find the ledge quite warm, while the panes of glass are comparatively cool. The radiation passes through the glass but is absorbed by the ledge.

The explanation is that not only the quantity but also the *quality* of the radiation from any source depends upon the temperature of the source. By saying this we do not mean to assert that there are different kinds of heat: but there are different kinds of heat-radiation. On these the glass exercises a selective absorption, stopping some of them while it lets others through. Thus it is opaque to the radiation from a source at a comparatively low temperature, whereas it transmits a considerable fraction of the radiation from a source at a high temperature.

## Examples on Chapters X–XII

1. Explain the use of the wire gauze surrounding the flame in the Davy lamp used in coal mines.

2. Describe how the heating of buildings, (*a*) by hot-water pipes, (*b*) by steam, depends on convection and conduction of heat, specific heat, and latent heat.

3. Should a kettle intended to be heated by standing in front of a fire be bright or black? State fully the reasons for your answer.

4. What are the differences in the behaviour of rock-salt, alum, and glass towards radiant heat? Would a rock-salt fire-screen prove efficient?

# CHAPTER XIII

## WIND, DEW, RAIN, ETC.

**93. Land- and Sea-Breezes.**—Winds are convection-currents produced by the unequal heating of the various parts of the surface of our globe. During the day-time the earth is warmed by the sun's rays and imparts its heat to the air lying above it. The sea has a lower absorbing power (Art. 90) than the earth, and its specific heat (p. 46) is higher: further, the evaporation from the surface tends to keep its temperature from rising. Thus the sea is cooler, during the day-time, than the land, and we should expect the colder and heavier air from the sea to flow inward, displacing the warmer and lighter air over the land (Fig. 65).

This is found to be the case near the sea-coast, especially on islands in tropical regions. In the morning a breeze from the sea springs up, increasing in strength towards the afternoon, and dying away at sunset. This is called the **sea-breeze**.

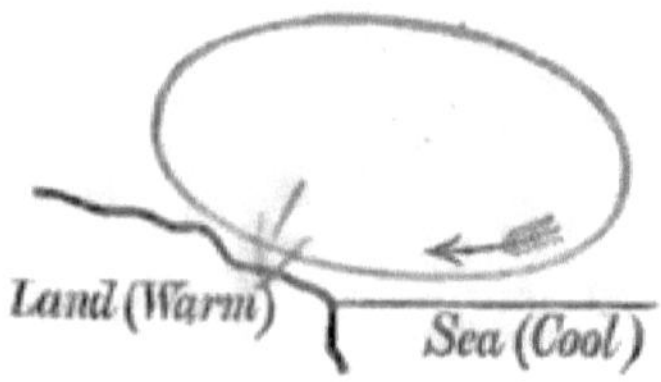

Fig. 65.—MORNING. 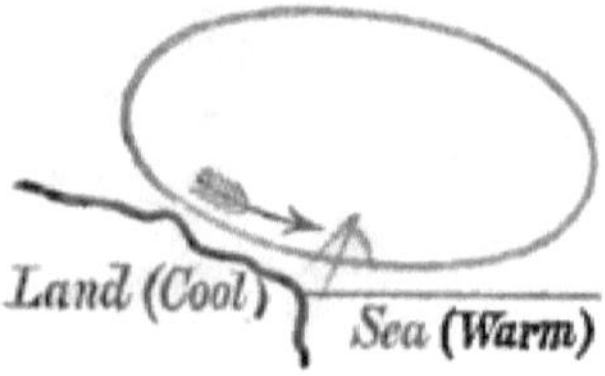   Fig. 66.—EVENING.

After sunset both land and sea are cooled by radiation. But the earth is a better radiator than the sea; it cools more rapidly; and so, during the night-time, the sea is warmer than the land. The warmer air above it rises upwards, being displaced by a colder current from the land (Fig. 66). This is the **land-breeze** which blows during the night and dies away towards morning.

**94. The Trade Winds.**—Of greater importance than these local winds is the great system of Trade Winds which blow all the year round over a belt extending about 30° on both sides of the equator. North of the equator the Trade is a N.E. wind (*i.e.* blows *from* the north-east); south of the equator it is a S.E. wind.

The Trade Winds are due to the intense heat of the sun in equatorial regions. The warmer and lighter air ascends and

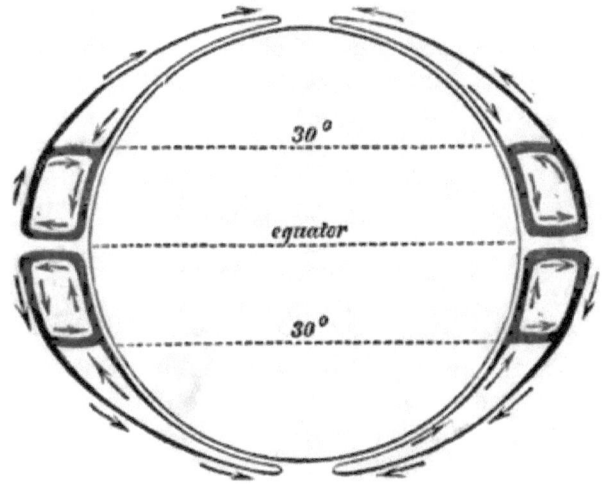

Fig. 67.

its place is taken by currents of air flowing in from colder latitudes both N. and S. of the equator (see Fig. 67). If this were all, we should expect to find in the northern hemisphere a Trade Wind blowing directly from the north, and in the southern hemisphere directly from the south. This would be the case if our earth were at rest. But it must be remembered that the earth rotates from west to east, and this eastward velocity *increases* as we pass from the poles to the equator. A mass of air moving in the nothern hemisphere from north to south only slowly acquires this motion towards the east: it is constantly moving southwards into regions which have a greater easterly velocity than itself. Hence it appears to be blowing towards the south-west; *i.e.* it is a N.E wind.

Similarly the Trade Wind in the southern hemisphere does not blow directly from the south but is a S.E. wind.

**95. Dew.**—On clear, cloudless nights the earth is cooled by radiation into space. The air in contact with it is also chilled, and if its temperature falls below the dew-point (Art. 63), a portion of the water-vapour present is deposited in the form of dew.

The deposition of dew is most copious *on calm, clear nights*. Clouds lessen the cooling effect of radiation; they reflect back the heat to the earth, so that its temperature does not fall as much on a cloudy night as when the sky is clear. Wind hinders the formation of dew, because it continually brings fresh portions of air into contact with the earth without allowing them to remain near it long enough to be cooled to the dew-point.

Again, dew is most copiously deposited upon substances which are *good radiators*, and which have a *clear view of the sky*. The substances on which dew is most readily formed are such as grass, plants, wool, and wood: these are good radiators and bad conductors of heat, so that they quickly cool on clear nights. Very little dew falls upon stones, slates, and metal surfaces: for although they may be fairly good radiators they are also good conductors; thus the loss of heat from the surface is made up for by conduction of heat from the interior or from the earth.

When the temperature of the air falls during the night below 0° the water-vapour present is deposited in the solid crystalline form as hoar-frost.

**96. Clouds, etc.**—Clouds consist of small particles of water produced by the condensation of the water-vapour contained in the air. Fogs and mists are simply clouds formed close to the surface of the earth.

Water-vapour itself is transparent and invisible like air. If you boil water in a kettle with a tightly-fitting lid and watch the steam coming out of the spout, you will see that there is a clear space for an inch or so in front of the spout. This clear space consists of invisible steam: the white cloud beyond is not steam at all but consists of small particles of liquid water. The steam is condensed partly by coming in contact with the colder atmosphere and partly by the cooling effect due to its own expansion. The steam inside the kettle is under a somewhat greater pressure than that of the atmosphere; and when it gets outside it expands. Now, whenever steam or air expands it does work as truly as if it were expanding in the cylinder of a steam-engine or air-engine: and, as a supply of heat is thereby taken from it (corresponding to the work done), it is cooled by its own expansion.

EXPT. 71.—On the plate of an air-pump place a well-fitting bell-jar and work the pump rapidly. After about half a dozen strokes the air inside the bell-jar generally becomes misty. By the rapid exhaustion (and expansion) the air is cooled, and, if the temperature falls below the dew-point, some of the water-vapour in it is condensed in the form of small liquid particles which form the mist. There is usually enough water-vapour present in the air to show this effect.

97. Rain, etc.—The particles of water forming a cloud tend to coalesce into larger drops, and, if these increase in size owing to further condensation, they presently fall as rain.

Snow is produced when the temperature of the air falls below 0°, and under the microscope is seen to consist of small hexagonal crystals (Fig. 68).

The causes that favour the fall of rain are similar to those by which clouds are formed, viz. the cooling of warm, moist air (1) by coming in contact with colder air or cold land, and (2) by its own expansion. This latter cause comes into play whenever

Fig. 68.—SNOW-CRYSTALS.

warm air rises upwards as a convection-current, or when it is forced to ascend, as in passing over a range of hills; for as the air rises it gets into regions where the pressure is less (Arts. 29 and 51), and therefore it expands.

In our own island the rainfall is heaviest on the western coasts, for the S.W. winds, having passed over the warm waters of the Gulf Stream, are themselves warm and moist: when they come in contact with the land, and especially when they have to pass over mountain-ranges, they become cooled and deposit their moisture as rain.

Rainfall is measured by means of an instrument called a **rain-gauge** (Fig. 69). This consists of a funnel for collecting the rain and a graduated vessel for measuring its volume. This volume, divided by the area of the funnel, gives the depth of the layer of water which would be produced if all the rain remained on the surface of the ground.

In Great Britain the average annual rainfall is somewhat under 30 inches. The driest district is Lincolnshire, where the rainfall is only about 20 inches. The rainfall is heaviest in the Lake District, the wettest place being the Stye in Borrowdale (Cumberland), where the rainfall is about 175 inches.

Fig. 60 —Rain-Gauge.

# ANSWERS TO EXAMPLES

### CHAPTER II (p. 13)

**1.** 7°·5 C.  **2.** 15° C.; 40° C.; 85° C.; 95° C.  **3.** 194° F.; 176° F.; 86° F.; 23° F.  **4.** −37°·3 F.; 674°·6 F

### CHAPTER III (p. 21)

**3.** 40·048 ft.  **4.** 1·224 ft.  **5.** 22·128 cm.  **6.** 2·0068 m.; 150°.  **7.** 0·0432 in.

### CHAPTER IV (p. 26)

**1.** 0·00002797.  **2.** 0·000302.  **3.** 0·000156i.  **4.** 131°·7.  **7.** 100·24 c.c.

### CHAPTER V (p. 36)

**4.** 320 c.c.  **5.** 273°.  **6.** 91°.  **7.** 12·38 litres; 13·2 litres.  **8.** 0·0036.  **9.** 546°.  **10.** 0·9462 gm.  **11.** Reduced to two-thirds.  **12.** 75 c.c.; 0° C.  **13.** 333°.  **14.** Volume remains unaltered.  **15.** 383·2 c.c.  **16.** 115·5 cub. in.  **17.** $\frac{1}{273}$.

### CHAPTER VI (p. 46)

**1.** 76,800 units.  **2.** 1995 units.  **3.** 3·1.  **4.** 0·2.  **5.** 88°·1.  **6.** 90°·9.  **7.** 17°.  **9.** 0·1327.  **10.** 853°.

## Chapter VII (p. 54)

**2.** 1250 gm. **3.** 80. **4.** 32 lbs. **5.** 80. **6.** 0·09. **8.** Ice is melted and the water raised to 20°. **9.** 0·1144. **10.** 1 lb. **11.** 173·7 pound-degrees-C.

## Chapter VIII (p. 70)

**4.** 31,200 units. **5.** 21,480 units. **7.** 8040 pound-degrees. **8.** 6·29 lbs. **9.** 536·3. **10.** 37·3 gm.

# LIGHT

1

# CHAPTER I

## INTRODUCTION—SHADOWS, ETC.

**1. Radiation of Light.**—Let us consider what happens when a body is gradually heated in a dark room. We may suppose the body to be the iron ball used in Ch. xii (Heat); or an iron wire made gradually hotter by passing an electric current along it.

At first, all that we are able to say is that, as the temperature increases, the amount of heat-radiation increases. But when the temperature reaches about 500°—the temperature known as 'a dull-red heat'—we begin to be aware of something else. The body becomes visible. In addition to the radiation which produces in our bodies the sensation of warmth, it now radiates out light which produces in our eyes the sensation of vision.

The colour of the light emitted is red. As the temperature increases the light becomes stronger and brighter: there is an increase in the amount of light radiated. At the same time the light becomes whiter—there is a change in the colour of the light radiated.

**2. Definitions.**—A body from which light proceeds is said to be luminous. If it emits the light of itself, it is said to be self-luminous. This is the case with bodies which are heated to a high temperature by their own combustion or otherwise. The sun is a self-luminous body; so also are candle-flames and gas-flames. The moon is not a self-luminous body; it simply reflects to our earth the light of the sun, just as sunlight is reflected by the surface of the sea or by a looking-glass.

Any substance through which light passes may be called a *medium*. If a medium possesses the same properties at all points and in all directions, it is called a *homogeneous medium*. Air, water, and glass are examples of homogeneous media.

A body through which objects can be seen distinctly (such as a sheet of glass) is said to be *transparent*. A body which allows light to pass through it, but through which objects cannot be seen distinctly, is said to be *translucent*. Thin paper, china, and ground glass are translucent. A body which does not allow light to pass through it at all is said to be *opaque*.

In general, when light falls upon any body, three things happen :—

(1) a portion of the light is reflected.
(2) a portion is transmitted.
(3) the rest is absorbed.

The surfaces of quicksilver and polished metals may be taken as typical instances of good reflectors. Air, water, and glass are good examples of transparent media; they transmit light with very little loss by absorption. Dull-black surfaces, such as those of velvet and lamp-black, are typical examples of good absorbers.

**3. Good Absorbers are Good Radiators.**—The following experiments show that the relation which we found to exist between absorbing and radiating powers for heat (p. 103) also holds good for light.

EXPT. 1.—Heat to redness in a crucible furnace a piece of earthenware having a strongly-marked pattern. The simplest plan is to choose a plate having a well-marked black or dark-blue pattern on a white ground, and to break this up into pieces which can be put into the crucible. The experiment must be performed in a darkened room.

With the aid of crucible tongs withdraw one of the pieces quickly and examine it before it cools. It exhibits a reversed pattern. Thus if the pattern of the (cold) plate seen in daylight be as in Fig. 1 (dark on a white ground), the pattern of the red-hot plate will be as in Fig. 2 (bright on a dark ground). The dark part—the good absorber—is also the good radiator.

Vary the experiment by making a chalk mark on an old poker : heat this in a fire and examine it in the dark.

EXPT. 2.—Mark a broad cross with ink on a piece of platinum-foil, and heat the foil in the flame of a Bunsen burner in a darkened room. (The ink used should not be aniline ink but common black ink made with

REVERSAL OF PATTERN.

Fig. 1 (Cold).      Fig. 2 (Hot).

copperas, which leaves a deposit of oxide of iron on the foil.) The face of the foil exhibits a *bright* cross on a *dark* ground. This shows that the iron oxide emits more light than the polished platinum.

[*Note.*—In the experiments to be described in this and the following chapters we shall sometimes use the sun as our source of light, the sunlight being admitted through a hole in the window-shutter of a darkened room; at other times the flame of a candle, lamp, or gas-burner, in which the light is produced by the burning of combustible bodies. In performing many experiments in light an optical lantern is indispensable, especially when a powerful light is required for showing an experiment to a number of persons. As a 'radiant' or source of light for lantern work an Argand gas-burner or a specially-constructed form of oil-lamp may be used; but the lime-light is at once more powerful and more convenient. A full description of the lantern and the methods of working with it would be out of place here: junior students are not likely to use it excepting under the direction of a teacher. The details of lantern manipulation are fully described in Mr. Lewis Wright's *Light.*]

§ 4. **Rectilinear Propagation of Light.**—Examine a beam of light as it passes through a darkened room when the sun is allowed to shine through a small round hole in the window-shutter. You can track its path as it passes through the room: it forms a divergent or conical beam. Strictly speaking, it is not the beam of light that you see. Light itself is not visible, and if there were nothing present to reflect the light to your eye you would not see the beam at all. But the air always contains fine particles of dust floating about in it.

These reflect the light to your eye and so enable you to follow the path of the beam.

Hold a sheet of paper in the path of the beam and at right angles to it: you see a bright round patch of light on the paper. This is an image of the sun. Its size increases as you move the paper farther away from the hole.

Everywhere the path of the light is straight: it does not bend round anywhere. Suppose straight lines to be drawn from the hole to all points on the circle of light which is thrown on the paper; the light travels along these lines. We may thus imagine the beam of light to consist of a pencil of rays, each ray being a straight line along which the light travels.

If you wish to test the matter further, you may take three cards and make a pin-hole in each. Pin the cards on wooden blocks, so that the holes are all at the same height and in a straight line. Place a candle in front of the first hole and look through the third. As long as the holes are in a straight line you can see the light shining through; but the light disappears as soon as any one of the cards is moved aside.

We shall find later on that a beam of light *is* bent when it passes from one medium to another (*e.g.* from air to water or glass), but so long as we keep to the same homogeneous medium we may say that *light travels in straight lines.*

5. **Images produced by Small Apertures.**—The round patch of light referred to in the last article is an inverted image of the sun. If there are brightly illuminated objects (*e.g.* houses in full sun-light) outside the dark room, inverted images of them may be produced in the same way by allowing the light from them to pass into the room through a small hole and on to a white screen placed behind. Fig. 3 shows how these inverted images are produced. Light is emitted in all

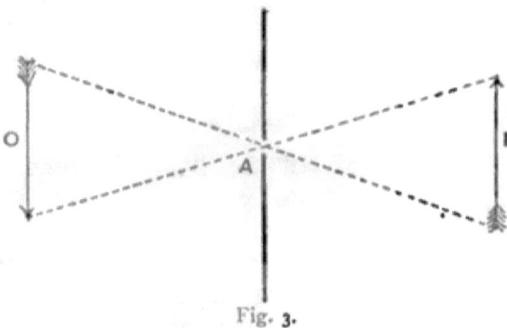

Fig. 3.

directions from every point of the bright object O. Of this, only a small pencil passes through the aperture A. This pencil produces upon the screen a small patch of light which is the image of the corresponding point of the object. Thus in the figure pencils of light pass in the direction of the dotted lines from the head and tail of the arrow and produce images at points on the screen. From points intermediate between the head and tail of O pencils of light proceed which produce images at corresponding intermediate points on the screen. Thus a complete image I is formed. It is also clear that whereas O is an arrow with its head down, I is an arrow with its head up. The image is an inverted one. Its size is directly proportional to the distance of the screen from the aperture. In the figure it is about the same size as the object. If the screen is brought nearer to the aperture the image becomes smaller; and if the screen is moved farther off it becomes larger. The images thus produced exhibit the natural colours of the objects, but are very faint, because the amount of light that can pass through a small hole is itself small. By making the hole larger a brighter image can be obtained, but it becomes blurred and indistinct at the same time.

EXPT. 3.—Remove the condensing lenses of a lantern and cover the front with a tin-foil cap. Make a pin-hole in this and place a paper screen in front. A faint image of the gas-flame (or other radiant in the lantern) is formed on the screen. Make other pin-holes near the first; each one produces a fresh image, but the images soon overlap and become confused. The same effect is produced by making one *large* hole.

EXPT. 4.—Make two tubes (A and B, Fig. 4) by wrapping pasted paper round a wooden cylinder. Blacken the insides with lamp-black varnish (p. 99) or line with black paper. One tube should slide within the other. Cover the end of the tube A with tin-foil and make a pin-hole in the middle of the foil. Cover the end of the other with tissue-paper. You have now a 'pin-hole camera,' which you can point towards a flame or bright object. The image can be seen by looking at the tissue-paper (*n*). It becomes larger and fainter as the tube B is drawn out.

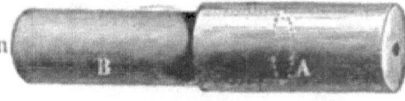

Fig. 4.

EXPT. 5.—For class demonstration the following is a good way of showing images produced by small apertures. Punch a clean hole (about

2 mm. diameter) in a large sheet of tin-plate or card-board and clamp this in a vertical position. On one side of it place a white screen and on the other side three candles arranged thus—.˙. It is then easily seen that the images occupy this relative position ˙.˙ and that the distance between increases as the screen is moved away from the hole.

**6. Shadows and Eclipses.**—The formation of shadows is a direct consequence of the propagation of light in straight lines.

When the source of light is a single bright point (S, Fig. 5) the form of the shadow thrown by any opaque object K is

Fig. 5.—Shadow Cone.

easily seen. For the cone of light which proceeds from S to K is stopped by the latter, and the portion of this diverging cone, which lies on the farther side of K is all in shadow. Sharp, well-defined shadows of this kind are thrown by electric lights (arc lights).

But when the source of light is of considerable size there is

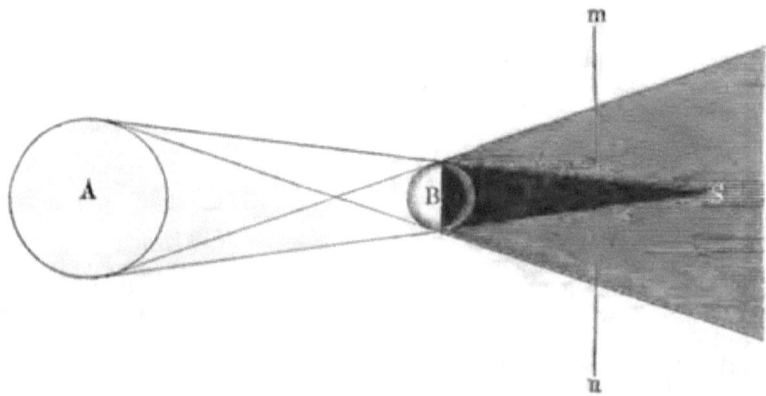

Fig. 6.—Umbral and Penumbral Cones.

produced, in addition to this total shadow or *umbra*, a half shadow or partial shadow called the *penumbra*. Suppose

A (Fig. 6) to represent the globe of a lamp and B an opaque obstacle, such as a ball or orange. Then it will be seen from the figure that the black cone is completely in shadow and receives no light from *any* part of A; this may be called the *umbral cone*. The shaded cone which diverges from a point between A and B is only partly in shadow, and may be called the *penumbral cone*. Every point in it receives light from *some* portion of the illuminating surface A. When a screen is placed behind B the shadow thrown upon it consists of a central umbra surrounded by a penumbra. The size of the latter increases as the screen is moved farther away from B; when the screen is in the position *mn* the shadow appears somewhat as shown in Fig. 7. The penumbra, however, is not of uniform depth all over. Its edge is not sharply defined: it gradually deepens from the outside inwards until it merges into the complete shadow of the umbra.

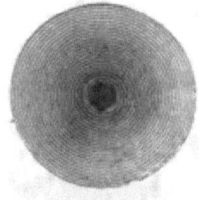

Fig. 7.

Fig. 6 also illustrates the way in which eclipses of the sun are produced when the moon comes between it and our earth. Suppose A to represent the sun, B the moon, and the screen *mn* a portion of the earth's surface. Within the umbral cone no part of the sun is visible; therefore to an observer stationed on any part of the earth within the umbra the sun will appear totally eclipsed. Within the penumbra a part of the sun is visible and the rest invisible, so that an observer stationed anywhere within the penumbra will see the sun partially eclipsed.

EXPT. 6.— The student should examine for himself the different kinds of shadows above described. For sharp shadows produced by luminous points use a lantern: remove the condensers and fit on a cap in which a small round hole has been pierced. Or, since the breadth of the penumbra increases with the distance from the object, any fairly small source of light (candle) will cast a sharp shadow on a screen placed near it. Farther off the shadow shows a penumbra.

Hold a pencil upright between a flat fish-tail burner and a wall. Examine the shadow. When the flame is 'edge on' the shadow is sharp and well-defined. When the flame is 'broadside on' the shadow is ill-defined (penumbra). In the first case the source of light is narrow, in the second case broad.

Examine the shadow of an orange or ball thrown on a paper screen by a lamp with a round ground-glass globe. Prick holes through the paper in various parts of the umbra and penumbra and look through them towards the lamp. Compare what you see with Fig. 6.

The penumbral cone is always divergent, as in the above figure. The umbral cone is also divergent when the object is larger than the source of light. It is convergent in Fig. 6 because B is smaller than A. When the source and the object are of the same size the umbral cone becomes a cylinder and the umbra is of the same size throughout.

### Examples on Chapter I

1. Explain exactly how a small hole is able to produce an image of an object on a screen. What will be the effect of enlarging the hole?

2. If an image produced by a pen-hole camera is half the size of the object, what are their relative distances from the hole? Why does the image become fainter as it becomes larger?

3. By means of a small hole in the window-shutter of a darkened room an image of a house 30 ft. away is thrown on a screen 6 in. from the hole. The image is 9 in. high : how high is the house?

4. If you hold a hair in sunlight close to a sheet of paper you can see the shadow of the hair on the paper; but if you hold the hair a couple of inches away from the paper you can scarcely see any trace of a shadow. How do you explain this?

5. When has an umbra a limited length? Under what conditions is the transverse section of the umbra cast by a body larger than the body itself?

6. A circular uniform source of light, 2 inches in diameter, is placed at a distance of 10 feet from a sphere 2 inches in diameter. Calculate, approximately, the diameters of the umbra and penumbra cast on a screen 5 feet beyond the sphere.

# CHAPTER II

## PHOTOMETRY

**7. Illuminating Power and Intensity of Illumination.**—It is evident that different sources of light give out different amounts of light. An ordinary gas-jet, for example, gives out ten or fifteen times as much light as a candle. This is expressed by saying that its illuminating power is ten or fifteen times as great as that of the candle. In this country the standard of illuminating power is the amount of light given out by a candle of a certain weight burning at a certain rate.[1] When we speak of a gas-flame as being of ten candle-power (10 C.P.) we mean that it gives out as much light as ten standard candles.

If such a flame were placed at a distance of a foot from a screen, it would throw upon the screen ten times as much light as a standard candle would when placed at the same distance. If the gas-flame were moved farther away, it would illuminate the screen less brightly. At a distance of four feet, for example, it would throw less light upon the screen than a candle would at a distance of one foot. We must, therefore, distinguish carefully between the illuminating power of a source of light and the intensity of illumination which it produces.

The **intensity of illumination** produced by a given source of light on a given surface is the quantity of light received per unit of that surface. This clearly depends upon two things—

(1) Upon the illuminating power of the source being

[1] 'Sperm candles of six to the pound, each burning 120 grains per hour.'

directly proportional to this; hence the intensities of illumination produced by different sources at the *same* distance are proportional to their illuminating powers.

(2) Upon the distance of the surface from the source of light.

**8. Law of Inverse Squares.**—It is evident that the intensity of illumination diminishes when the distance increases. But it would not be correct to say that the intensity decreases *as* the distance increases (*i.e.* in the same proportion). For example, when the distance is doubled the intensity is reduced, not to one-half, but to one-fourth. This may be shown both by experiment and calculation.

EXPT. 7.—Make a small square frame of wire and fix it, as at O in Fig. 8, half-way between a candle and the wall of a

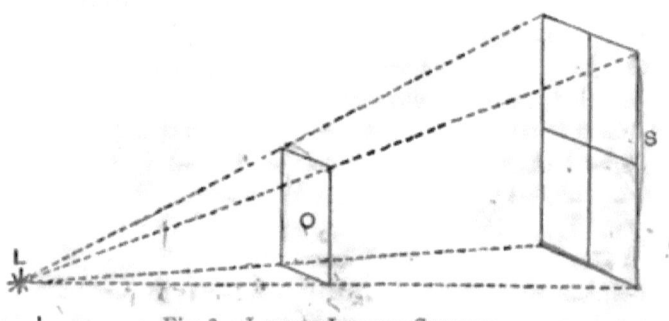

Fig. 8.—LAW OF INVERSE SQUARES.

darkened room. It throws a square shadow upon the wall. Cut a piece of paper to the size of this shadow (S) and fold it across and across as shown by the lines. The folded sheet just fits the frame O. S is divided into four squares each equal to O; its area is four times that of O.

If O were an opaque obstacle, it would block out all the light from the space S. The same amount of light which falls upon O is at *twice* the distance spread out over S, which has *four* times the area. Clearly then the intensity of illumination (or amount of light per unit surface) at S is only *one-fourth* of that at O.

Again, if O were placed at one-third of the distance from the candle to the wall, the area of S would be nine times that of O (S being three times as large each way). In this case S would be *three* times as far off from L as O is, and the intensity of its illumination would be *one-ninth* of that at O.

Similarly, if O were placed at *one-quarter* the distance from L to S, the illumination at S would be *one-sixteenth* of that at O.

All this depends upon the fact that light travels in straight lines; from which it follows that the areas of perpendicular sections of the beam at different distances are proportional to the squares of these distances.

Now observe that $\frac{1}{4} = \frac{1}{2^2}$; $\frac{1}{9} = \frac{1}{3^2}$; $\frac{1}{16} = \frac{1}{4^2}$.

We may therefore express the result according to the following law of inverse squares :—

The intensity of the illumination produced by a source of light is *inversely proportional to the square of the distance from the source.*

9. **The Shadow Photometer (Rumford's).**— Photometry deals with the comparison and measurement of the illuminating powers of different sources of light. One of the simplest methods of doing this is by means of the shadow photometer introduced by Count Rumford. In the following experiment we shall describe how to use the shadow photometer for comparing the illuminating powers of a lamp and a candle. If the candle used be a standard candle (p. 123), our experiment will enable us to measure the 'candle-power' of the lamp.

EXPT. 8.— Place a vertical rod (a ruler, or the rod of a retort-stand) in front of a white screen. Begin by putting the candle and lamp side by side, about a foot in front of the rod. Two shadows of the rod are thrown on the screen, one by the candle and one by the lamp, and it is evident that the latter is much the darker of the two. What does this mean? Remember that you have now two sources of light illuminating the screen. The part of the screen on which the candle-shadow falls receives no light from the candle, but it *does* receive light from the lamp. Similarly, the part of the screen on which the lamp-shadow falls receives no light from the lamp, but it *does* receive light from the candle. Since the former shadow is not so dark as the latter, and both sources of light are at the same distance, we conclude that the lamp gives out more light than the candle.

Now move the lamp farther away, until the shadows appear of the same depth. The lamp should be placed in such a position that the two shadows are formed side by side, as in Fig. 9, so that you can easily compare their intensities.

Measure off with a foot-rule or metre-scale the distances from the screen to the candle and from the screen to the lamp. We shall suppose these to be 1 ft. and 2 ft. respectively

Fig. 9.—SHADOW PHOTOMETER.

when the shadows are of equal depth. Then it follows that the illuminating power of the lamp is to that of the candle as $(2)^2$ is to 1, or as 4 is to 1.

For if we denote by P the illuminating power of the lamp, then, as explained in the preceding article, the intensity of the illumination which it produces at a distance of 2 ft. is proportional to $\frac{P}{(2)^2}$ or $\frac{P}{4}$. In the same way the intensity of the illumination produced by the candle (whose illuminating power is 1) at a distance of 1 ft. is $\frac{1}{(1)^2}$ or 1. Since these are equal (the shadows being equally illuminated) we have

$$\frac{P}{4} = 1, \text{ or } P = 4.$$

Hence the lamp is of four candle-power.

Or you may look at it in this way. In their present positions the candle and lamp illuminate the screen with equal intensity.

If you were to move the lamp to where the candle is, you would be diminishing its distance from the screen from 2 ft. to 1 ft. This would *increase* the intensity of the illumination on the screen due to it in the proportion of $1^2$ to $2^2$, or 1 to 4. Thus the intensity of the illumination due to the lamp is four times as great as that due to a candle at the same distance; in other words, the lamp gives out as much light as four candles.

In connection with shadow photometers remember these two points—

(1) Although you make the comparison by judging of the comparative depths of two shadows, you are really comparing the intensities of illumination on two adjacent portions of the screen and adjusting these to equality.

(2) The illuminating powers of the two sources of light are *directly* proportional to the squares of their distances from the screen. If the distances had been 1 ft. and 3 ft. (instead of 1 ft. and 2 ft.), the illuminating powers would have been as 1 to 9, and so on. This also holds good for the photometer described in the next article. [Compare this statement carefully with Art. 8.]

**10. The Bunsen Photometer.**—If you make a grease-spot on a sheet of paper and hold it up between you and the light, the spot appears brighter than the rest of the paper. Greasing the paper makes it more transparent and so more light gets through the part that has been greased. But if you stand with your back to the light and hold the sheet so that the light falls on it, the grease-spot appears darker than the rest of the sheet. The reason is obvious. The light that now reaches your eye from the sheet is reflected light. But, since the grease-spot lets through more light than the rest of the paper does, it cannot reflect as much light to your eye, hence it appears darker by contrast. A form of photometer depending upon these principles was introduced by Bunsen. It is simple in construction, and is in everyday use for the purpose of testing the illuminating power of coal-gas, etc.

EXPT. 9.—*To make and use a Bunsen photometer.*—Drop some melted wax from a paraffin candle on a piece of stoutish white paper (unglazed). Remove the excess with a knife when it has solidified. Press a hot flat-iron or spatula on the paper, so as to melt the wax well into it. The spot should be about the size of a halfpenny. Mount the paper disc as flat as possible in a suitable frame (Fig. 10).

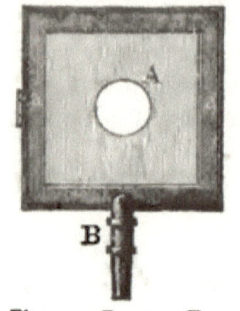

Fig. 10.—Bunsen Disc.

Draw a long straight chalk-line (or stretch a strip of paper) on a table, and on this mark a scale of centimetres or inches. At one end of the scale place the candle, and at the other end the lamp which is to be compared with it, blocking the one or the other up until the flames are at the same height. Between them place the disc with the grease-spot at the same height as the two flames. Looked at from the side facing the lamp, the grease-spot appears darker than the rest of the disc. Looked at from the side facing the candle, it appears brighter; this is the appearance which it always presents when the back of the disc is more brightly illuminated than the front.

Now move the candle up towards the disc: at a certain point the bright spot in the centre disappears, or very nearly so. It reappears as a dark spot if you overshoot the mark. Adjust the position of the candle so that the spot is as nearly as possible invisible on both sides of the disc. When this is the case the disc is equally illuminated on both sides. The grease-spot appears neither darker nor brighter than the rest of the disc, because the amount of light that is lost by passing through it is exactly made up for by the transmission of an equal amount of light from the opposite side.

Measure off the distances from the disc to the candle and the lamp. Their illuminating powers are directly proportional to these distances (see Art. 9). Suppose the distances to be 11 in. and 55 in. These are as 1 to 5. Their squares are as 1 to 25. The lamp is, therefore, of 25 candle-power.

EXPT. 10.—*To verify the law of inverse squares.*—Place four candles in a row across the line drawn on the table, facing the Bunsen disc, and 2 or 3 ft. from it. On the other side of the disc place a single candle from the same batch. Adjust the position of the single candle or the disc until the latter is equally illuminated on both sides (*i.e.* until the grease-spot disappears). When this is the case it will be found that the row of four candles is twice as far from the disc as the single candle is. What conclusion do you draw from this?

**11. Note on Optical Benches, etc.**—In working with the Bunsen photometer, as well as in many other optical experiments (*e.g.* in finding focal lengths of mirrors and lenses), it is convenient to have the various pieces of apparatus (candles, screens, etc.) mounted on suitable holders which slide on a graduated bar. Such an arrangement is called an **optical bench**. A short portion of a simple form of wooden bench is shown in Fig. 11, together with one of the sliding pieces. Each of these is furnished with an upright socket consisting of a piece of brass tube about 5 in. high and $\frac{1}{4}$ in. in diameter. The mirrors, lenses, etc., are mounted in wood or cork mounts, attached to iron rods which fit into the sockets; if these are pro-

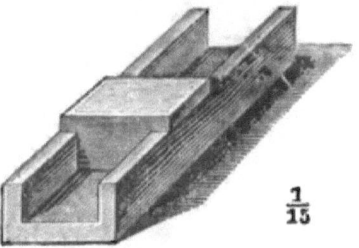

Fig. 11.

vided with screws, the rods can be clamped at any convenient height. The bench should be about 6 ft. long, and a strip of stout white paper should be carefully pasted along the side of it. When this has dried thoroughly mark a graduated scale on it, the graduations coming right up to the edges of the slides. A centimetre scale is much more convenient than one in feet and inches. On the edge of each slide make a vertical mark, corresponding to the centre of its socket.

The measurements required may also be made by using a beam-compass (Fig. 12), or even by holding a graduated scale alongside the apparatus.

Fig. 12.—Beam-Compass.

But it is a great convenience to have some form of optical bench, especially when pieces of apparatus have to be moved backwards and forwards and always kept in the same straight line.

### Examples on Chapter II

1. A candle is placed at a distance of 1 ft. from a cardboard screen, and a lamp of nine candle-power is placed at a distance of 12 ft. on the other side. Compare the illumination on the two sides of the screen.

K

The intensity of the illumination on the side facing the candle is to that on the side facing the lamp as $\frac{1}{(1)^2}$ to $\frac{9}{(12)^2}$, or as $1$ to $\frac{9}{144}$; *i.e.* as 144 to 9, or as 16 to 1.

2. How near to the screen would the lamp in the last question have to be moved in order to reverse the proportion, *i.e.* to make the intensities of illumination as 1 to 16?

3. A standard candle and a gas-flame are placed 6 ft. apart, the gas-flame being of four candle-power. Where would a Bunsen disc have to be placed between them so as to make the grease-spot disappear?

Let $x$ denote the distance of the disc from the candle; its distance from the gas-flame will then be $6-x$. The grease-spot disappears when the disc receives equal illumination from the candle and the lamp. When this is the case we must have

$$\frac{1}{x^2} = \frac{4}{(6-x)^2},$$

or, taking square roots,

$$\frac{1}{x} = \frac{2}{6-x}.$$
$$\therefore 6-x = 2x,$$
and
$$x = 2.$$

The disc must therefore be placed 2 ft. from the candle and 4 ft. from the lamp.

4. What do you mean by the intensity of the illumination at a point, and how would you show, experimentally, that it is inversely proportional to the square of the distance of the point from the source?

5. In testing a night-light against a candle by means of a Rumford photometer it is found that the shadows are of equal depth when the night-light is 2 ft. from the screen and the candle 3 ft. Compare their illuminating powers.

6. Two equal sources of light are placed on opposite sides of a disc, one being 20 cm. from it and the other 30 cm. Compare the intensity of illumination on the two sides of the disc.

7. In measuring the illuminating power of a gas-flame by a Bunsen photometer, the distance from the gas-flame to the grease-spot was 96 cm., and from this to the standard candle 30 cm. What was the candle-power of the gas-flame?

8. A Carcel lamp of nine candle-power is placed at a distance of 4 ft. from a standard candle: find, as in Example 3, the position in which a screen must be placed between them so as to receive equal amounts of light from each.

# CHAPTER III

## VELOCITY OF LIGHT

**12.** When a gun at some distance from you is fired, you see the flash before you hear the report. The sound takes an appreciable time to travel from the gun to you. But how about the light? Is the flash seen at the very same instant when it is produced? or does the light travel with a definite velocity? If it does, by what method can this velocity be measured?

Suppose a gun to be fired at regular intervals, say every hour. An observer near at hand and provided with a delicate chronometer observes the times at which the flashes are seen. Now suppose the observer to move a very great distance off. *If the distance were great enough*, he would not now see the flashes exactly at the actual times of firing (*i.e.* at the hour), but a short time afterwards; and by carefully measuring this time, knowing his distance from the gun, he would be able to calculate the speed with which light travels. As a matter of fact, this speed is so great that it would be exceedingly difficult to detect any retardation within any practicable distance on our earth. Astronomers have to deal with immensely greater distances, and so it is not surprising to find that the first observations on the velocity of light were made by astronomical methods.

**13. Römer's Method.**—The first estimate of the velocity of light was made about the year 1675 by a Danish astronomer named Römer. It happens that one of Jupiter's satellites (or moons) passes into the shadow of the planet at regular intervals ($48\frac{1}{2}$ hours), and is thus eclipsed. While the earth is in the neighbourhood of E, Fig. 13 (*i.e.* nearest to Jupiter), its distance from Jupiter does not change rapidly, and the successive eclipses of the satellites are seen to occur at regular and equal intervals. The same holds good when the earth is

Fig. 13.

in the neighbourhood of E' (farthest from Jupiter). But between these two positions, as the earth moves from E to E', its distance from the planet continually increases, and Römer observed that during this time the intervals between successive eclipses were always longer than the time mentioned. An eclipse seen at E' is 16 minutes 26 seconds later than if it had been observed at E. The difference is due to the time taken by the light in travelling from E to E', *i.e.* across the diameter of earth's orbit. This is a distance of about 184,000,000 miles, and since light traverses it in 986 seconds we conclude that its velocity is $\frac{184,000,000}{986}$, or about 186,600 miles per second. Some idea of this may be obtained by considering how long light would take to travel round our earth. The circumference of the earth is less than 25,000 miles. Light would therefore travel round it seven and a half times in a second.

**14. Fizeau's Method.**—It might appear impossible to measure such an enormous velocity as this otherwise than by astronomical observations. The skill of modern experimenters has, however, proved equal to the task. The first experimental method was devised by Fizeau, and depends upon the following principle.

If a toothed wheel be made to revolve very rapidly about its axis, the time taken by a single tooth or a single space in passing a given point is extremely small. For example, it is easy to make it as small as the ten-thousandth part of a second; during which time light would only travel about $18\frac{1}{2}$ miles. Now suppose that a beam of light passes between the two teeth of the mirror, parallel to its axis, and falls upon a distant mirror which reflects the beam back along its own path. When the light gets back again to the wheel it may be able to pass through a space, or it may be blocked by one of the teeth; and which of these two things happens will simply depend upon the distance of the mirror and the speed of rotation of the wheel.

Fizeau chose two stations 8633 metres apart. At one of these was placed the toothed wheel $ww$ (Fig. 14), and a bright source of light; at the other the mirror $m'$, which reflected the light back to the first station. The source of light was not

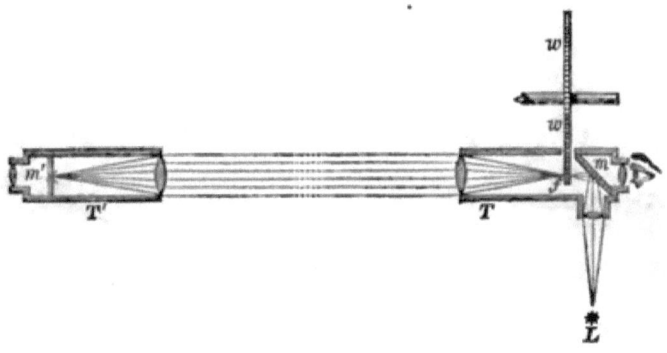

Fig. 14.—Fizeau's Method.

placed directly behind the wheel but to one side, at L, and the light from it was reflected to the distant mirror $m'$ by means of a transparent mirror $m$ (of unsilvered glass) inclined at an angle of 45°. Thus an observer looking through $m$ towards the distant station would see a bright star or point of light—the image of L produced by reflection in $m'$.

Now it was found that when the wheel was made to rotate at a certain speed this bright star was eclipsed; and the speed of rotation at which this happened was such that the wheel only took $\frac{1}{18,144}$ of a second to move through the breadth of a tooth or space. During this short time the light had travelled through one of the spaces at $f$ to the distant mirror $m'$ and back again, or through a distance of $2 \times 8633$, or

17,266 metres; and on its return its path was blocked by the next tooth. From this we conclude that in one second light would travel a distance of $17,266 \times 18,144 = 313,274,304$ metres. When the wheel was made to rotate twice as rapidly the star reappeared, the light now passing through a space at $f$ on its way to $m'$, and through the *next* space on its return journey.

Later experiments have shown that Fizeau's result was too high, and that the velocity of light is probably about 300 million metres per second, or 186,400 miles per second.

# CHAPTER IV

## REFLECTION OF LIGHT—PLANE MIRRORS

**15. Regular and Irregular Reflection.**—Light may be reflected from the surfaces of bodies either regularly or irregularly. Light is regularly reflected from smooth and regular surfaces, such as those of glass and polished metals; also from the surfaces of still water and mercury. Irregular reflection is more common. Even surfaces which appear smooth (*e.g.* paper, cardboard, etc.) are really full of small irregularities: consequently they diffuse the light that falls upon them, reflecting it irregularly in all directions.

EXPT. 11.—Take into a darkened room an ordinary mirror (looking-glass), a sheet of tin-plate, and sheets of white and black paper or cardboard. Admit a small beam of sunlight into the room (or use a beam from the lantern). Hold the mirror in the beam: it reflects the beam in a definite direction and throws on the wall a distinct patch of light, which moves about as you move the mirror. Try the tin-plate. The patch of light which it throws is not so bright or distinct: it is not such a good reflector. The white paper does not give any distinct patch of light; still it does in some way act as a reflector. This you can show by placing a white screen near the beam (but not in it) and holding the paper so as to throw the light on the screen. The screen becomes more or less uniformly lit up with diffused light.

Now get some one to hold the mirror again in the beam, while you stand in a far part of the room and look at it. If it is held so as to reflect the beam directly towards you, you see a glare of light; otherwise you can scarcely see the mirror at all (especially if the room is well darkened). The white paper, on the other hand, can be seen almost equally well from any part of the room: it reflects the light in all directions, so that some always reaches your eye. The black paper again is almost invisible. This is because it absorbs all the light that falls on it, so that none is reflected to your eye, whereas in the case of the mirror it was because all the light was reflected in another direction.

**16. Reflection from Plane Mirrors.**—When we speak of light as being reflected regularly, we mean that it is reflected according to some rule. We have now to find what this rule or law is.

EXPT. 12.—Cover the front of the lantern with a blackened cap in which has been cut a horizontal slit. Allow a beam of light [1] from this to fall upon a mirror facing the lantern and

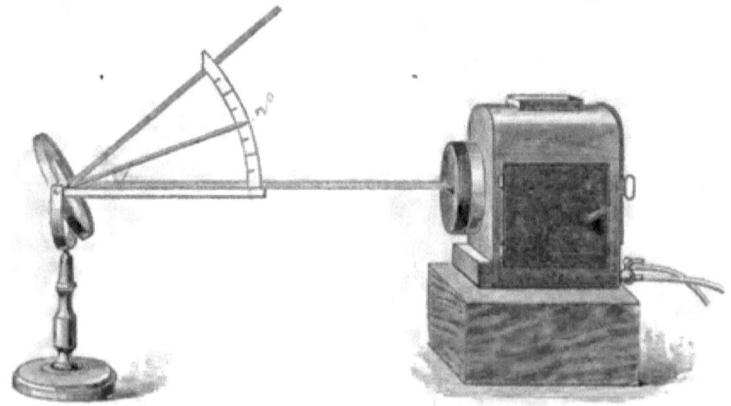

Fig. 15.—REFLECTION OF LIGHT.

fitted up as shown in Fig. 15. The mirror can be rotated about a horizontal axis. Attached to the mirror, and perpendicular to it, there is a light pointer which moves over a graduated arc fixed to a horizontal arm. The angle through which the pointer moves (reckoned from the zero at the end

---

[1] A beam of sunlight, reflected into the room from a mirror suitably placed outside, will do equally well. The room must, of course, be darkened. If a lantern is used, the objective must be removed and the rays rendered parallel by placing the source of light at the principal focus of the condenser.

of the horizontal arm) indicates the angle through which the mirror is moved out of the vertical. The height of the lantern should be adjusted so that the horizontal beam falls on the axis of rotation of the mirror. The path of the beam can be shown up more clearly by burning touch-paper[1] underneath (Art. 4).

Place the mirror vertically; the pointer stands at the 0° mark. The beam of light falls perpendicularly on the mirror *and is reflected back along its own path.*

Tilt the mirror up so that the pointer stands at (say) 20°; the reflected beam makes an equal angle with it on the other side (*i.e.* stands at 40°). Try other angles. Observe that whenever the mirror is rotated through any angle, *the reflected beam moves through double the angle; and the pointer always bisects the angle between the incident and reflected beams.*

**17. Definitions—Laws of Reflection.**—Suppose a ray of light IN (Fig. 16) falling on a plane reflecting surface at N to be reflected along NR. The line NP, drawn perpendicular to the surface at the point of incidence N, is called the *normal* to the surface. The angle INP, between the incident ray IN and the normal, is called the *angle of incidence.* The angle PNR, between the reflected ray and the normal, is called the *angle of reflection.*

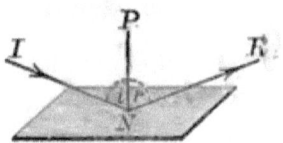

Fig. 16.

Having defined these terms, we may now state the laws of reflection as follows:—

I. *The reflected ray lies in the plane containing the incident ray and the normal, and on the opposite side of the normal.*

II. *The angles of incidence and reflection are equal.*

When a ray of light is incident normally on a mirror, the angles of incidence and of reflection are both zero, and the ray is reflected back along its own path.

In general, if you wish to find the path of the reflected ray, draw the normal at the point of incidence: on the other side

---

[1] Touch-paper is made by soaking brown paper in a solution of saltpetre and then allowing it to dry.

of the normal draw a line making with it an angle equal to the angle of incidence. This is the direction of the reflected ray.

Fig. 17 illustrates the nature of what was referred to in Art. 15 as irregular reflection. Every ray in the parallel beam which falls upon the unpolished surface is really reflected according to the usual law: but, as the various parts of the surface are inclined at different angles to the incident rays, the light is reflected in all directions, or *scattered*.

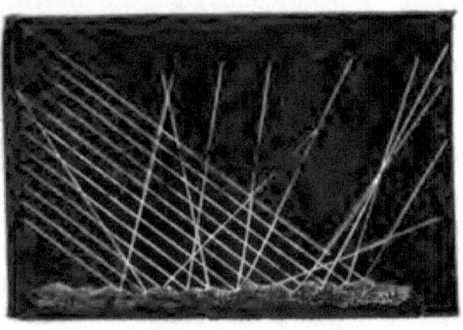

Fig. 17.—SCATTERING OF LIGHT.

**18. Experimental Verification.**—The apparatus illustrated in Fig. 18 is easily made, and by means of it the student can verify with considerable accuracy the relation between the angles of incidence and reflection.

EXPT. 13.—Cut a semicircle of about 1 ft. radius out of a wooden board or piece of stiff cardboard, and paste white

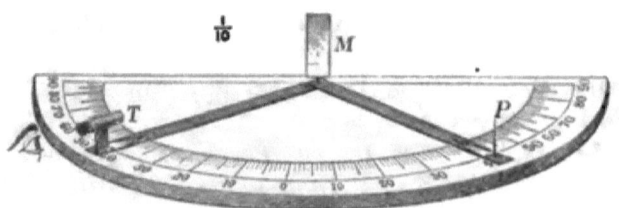

Fig. 18.—VERIFICATION OF LAW OF REFLECTION.

paper on it. When dry, mark off a semicircle on it, and divide it into degrees as in the figure. The movable arms MP and MT are strips of thin cardboard in which are cut V-shaped slits, the points of the V's lying just over the marks on the graduated circle. On one of the arms is mounted a tube T made of black paper (or a glass tube blackened on the inside), and a thread is stretched vertically across this tube.

The other arm carries a pin P. M is a mirror (2 in. by 1 in.) fastened by sealing-wax to a stout pin which is driven in, through the ends of the two arms, at the centre of the semi-circle. The mirror must be adjusted so that it is vertical and faces the 0° mark.

Place the arm P so that the end of the slit in it lies exactly over one of the degree marks. Move the arm T until the image of the pin in the mirror can be seen through the tube just behind the thread. Read off the angles.

**19. Measurement of Small Deflections.**—We may here refer to a very useful application of the law of reflection in measuring small angles, *e.g.* in observing and measuring small deflections of a suspended magnet. The method consists in attaching a small mirror to the magnet, and allowing a beam of light to fall normally upon this. If now the magnet moves through any angle, the mirror moves with it, *and the reflected beam moves through double the angle.*

This we have already proved in Expt. 12. It also follows from the law of reflection. For let *st* (Fig. 19) be the position of the magnet at rest and *oa* a ray of light falling normally upon the mirror attached to it at *a*; in this position the ray of light is reflected back along its own path. If now the magnet moves through an angle V into the position *s't'*, the normal to the mirror moves through the same angle into the position *ap*. The angle of incidence is *oap* = V, and hence the angle of reflection must = V; thus the reflected ray travels along *ax*, making an angle 2V with its original direction (*ao*).

The reflected ray thus forms a long weightless pointer which rotates twice as fast as the mirror. Its position is observed by allowing it to fall upon a graduated scale *mn* and watching the motion of the spot of light. By placing the scale at a sufficient distance from the mirror, the motion of the latter may be magnified as much as we please.

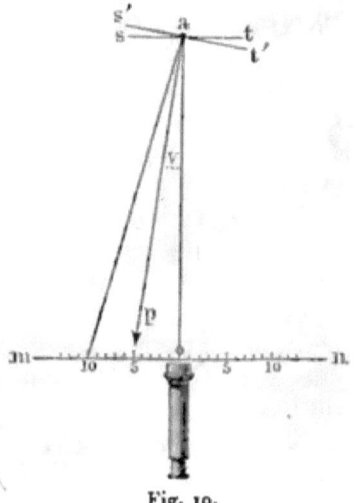

Fig. 19.

**20. Images in Plane Mirrors.**—Let A (Fig. 20) be a luminous point and MM a plane mirror. Rays of light proceed from A in all directions; some of these fall upon the mirror and are reflected by it. Let AB be any one of these rays. At B draw the normal BC; also draw BD, making the angle DBC equal to the angle ABC. Then by the law of reflection (Art. 17) BD is the reflected ray. To an observer whose eye is placed anywhere along BD the light would appear to come from some point behind the mirror in the direction of DB produced. From A draw a perpendicular to the mirror and produce it to meet DB produced at A'.

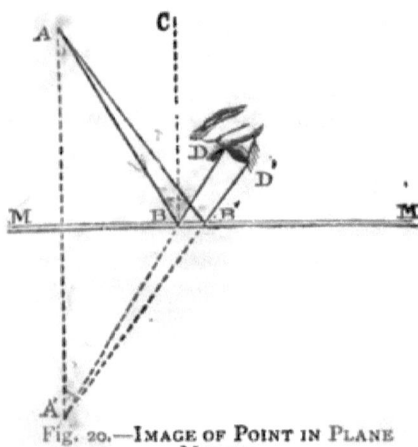

Fig. 20.—IMAGE OF POINT IN PLANE MIRROR.

Now, since AA' and CB are parallel (both being perpendicular to the mirror), the angles CBA and BAM are equal; and so also are the angles CBD and BA'M. But the angles CBA and CBD are, by the law of reflection, equal; therefore the angle BAM is equal to the angle BA'M. Thus in the triangles AMB and A'MB the base is common, the angles at A and A' are equal, and the angles at M are right angles. Hence the two triangles are equal, and the side A'M is equal to AM.

A' is as far behind the mirror as A is in front of it. Hence A' is a fixed point. We have supposed AB to be *any* one of the rays falling on the mirror (and lying in the plane of the paper). Hence any other ray such as AB' will also after reflection appear to come from the same point A'; it will be reflected along B'D', and the line D'B' produced backwards will pass through A'.

All rays diverging from A appear, after reflection in the mirror, to diverge from A'. A' is called the image of A. An observer looking at the mirror sees the image by means of a small pencil of these divergent rays. The position of the image does not depend upon the position of the observer's eye. It is not necessary that the mirror should extend right up to (or opposite) the object A; thus in the figure the portion BB' of the mirror alone is required in order that the observer may see the image A'.

The reflected rays do not really proceed from A' but only appear to do so; A' is therefore called an apparent or *virtual image*.

We have thus shown that **a pencil of** divergent rays falling on a plane mirror remains divergent after reflection. It will be a useful exercise for you to draw a diagram showing that a pencil of convergent rays remains convergent after reflection, also that a beam of parallel rays remains parallel after reflection.

**21. Summary and Definitions.**—*The image of a point in a mirror is a corresponding point from which rays of light diverge (or appear to diverge) after reflection from the mirror.*

*If the rays of light after reflection from the mirror really diverge from the point, it is called a real image; but if they only appear to diverge from the point, it is called a virtual image.*

*The position of the image of a point in a plane mirror is found by drawing a perpendicular from the point to the mirror and producing it until its length is doubled.*

**22.** We can now easily find by geometrical construction the position and size of the image of an extended object produced in a plane mirror.

Let AB (Fig. 21) be the object and MM' the mirror. From A draw a perpendicular to the mirror and produce it until its length is doubled. The point A' thus found is the image of A. In the same way the image of B is formed at B', which is as far behind the mirror as B is in front of it. Similarly the images of points intermediate between A and B are formed at corresponding points between A' and B'. Thus a complete image A'B' of AB is obtained.

From every point on the object rays of light proceed towards the mirror and, after reflection, appear to come from the corresponding point of the image. In Fig. 21 are shown the paths of the pencils of light by which an observer sees the extreme points of the image. When the observer's eye is in the position shown, the only part of the mirror required for the formation of the image is that lying between C and D.

The image is virtual and erect: it appears to be as far behind the mirror as the object is in front of it, and is of the same size and shape as the object.

But the appearance of the image is not exactly the same as that of the object facing the mirror. It is affected by what is known as *lateral inversion*. You will best understand what this is by placing a printed page in front of a looking-glass and trying to read the print in the image. The letters are all erect,

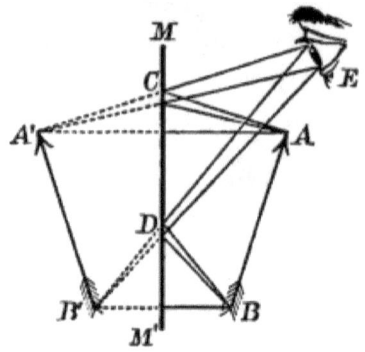

Fig. 21.—Image in Plane Mirror.

Fig. 22.—Lateral Inversion.

but the words read from right to left. A letter p appears in the image as a letter q (Fig. 22). A page of type, as set up by a printer, can be read by holding it in front of a mirror.

The effects of this lateral inversion can frequently be observed by looking at the face of a friend in a mirror. Our faces are never perfectly symmetrical, and so we do not quite see ourselves as others see us.

**23. Multiple Images in Inclined Mirrors**—Expt. 14.—Take two mirrors (about 4 inches square) and join two of their edges by pasting on a strip of cloth or ribbon, thus making a hinge about which the mirrors can be moved. Stand them vertically on a table with a lighted candle-end between. A number of images of the candle-flame are seen.

Either of the mirrors used by itself would give only *one* image. The effect observed must be due to a joint action of the mirrors, by which the light is repeatedly reflected from one mirror to the other and finally to your eye. These multiple reflections give rise to multiple images.

The number of images seen depends upon the angle between the mirrors. Adjust them so that this angle is 45°: the number of images is seven. Gradually increase the angle to 60°: the middle image disappears and the ones on each side of it (*i.e.* the two nearest the hinge) coincide and form a single image. The number is now five. Increase the angle to 90°: the same process is repeated and the number of images is reduced to three. Observe that the coincidence takes place when the angle between the mirrors is contained an exact number of times in four right angles (360°); and that if this number be $n$ the number of images is $(n-1)$.

It will be a useful exercise for the student to trace the paths of the rays by which these images are seen. In doing so the following principles should be borne in mind. When rays of light from a luminous object are reflected by a plane mirror, they diverge from a corresponding point at an equal distance behind the mirror: and this may be called the *primary image* of the object, the rays forming it having been reflected once only. If these rays fall upon a second mirror they will, after a second reflection, appear to diverge from a point as far behind the second mirror as the primary image, is in front of it. In other words, a *secondary image* of the object is formed by successive reflections of rays at the *two* mirrors. In finding the position of this secondary image, the primary image may be treated as a virtual object sending out rays of light.

This process may be repeated any number of times as long as any one of the images lies in front of one of the mirrors (*i.e.* in front of the plane in which the mirror lies). Thus a secondary image may after a *third* reflection give rise to a *tertiary image*, and so on.

Fig. 23 shows the appearance presented when a candle is held between two mirrors inclined at right angles. Two of the images seen are simply

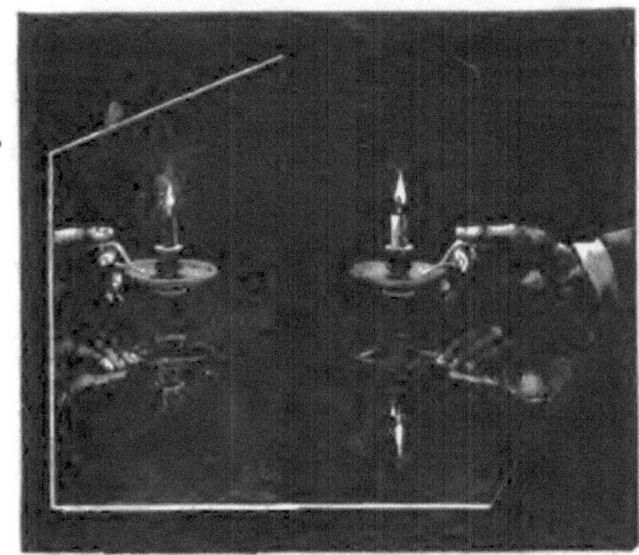

Fig. 23.

primary images produced by single reflections from the horizontal and vertical mirrors respectively. The third is a secondary image, and the paths of the rays by which it is seen are shown in Fig. 24. Rays proceeding from the object O and falling upon the horizontal mirror CM appear, after reflection, to proceed from the primary image $I_1$. Some of these rays fall upon the vertical mirror CM'. $I_1$ lies in front of this mirror, and may be regarded as a virtual object sending out rays of light which, after a second reflection (from CM'), appear to diverge from the point $I_2$, which is as far behind the vertical mirror as $I_1$ is in front of it. $I_2$ is a secondary image of O produced by successive reflections from the horizontal and vertical mirrors.

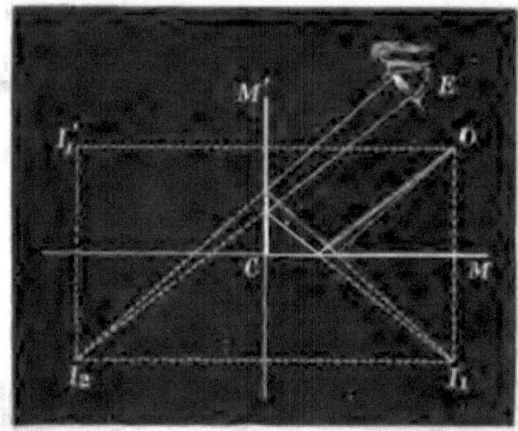

Fig. 24.

$I_2$ may be regarded as consisting of two coincident secondary images. For by a single reflection at the vertical mirror a primary image is produced at $I_1'$. This is in front of the horizontal mirror, and hence, as already described, a secondary image would be produced at a point as far behind the horizontal mirror as $I_1'$ is in front of it. But it is clear from the figure that this secondary image would coincide in position with $I_2$.

**24. Parallel Mirrors**—EXPT. 15.—Stand two mirrors vertically on a table and facing one another. Between them place a candle or other object. Look over the edge of one mirror into the other. You see a large number of images, and if the mirrors are exactly parallel, these images lie all in a straight line. They are produced by repeated reflections of light from one mirror to the other until the light reaches the eye. Good examples of these multiple images can often be seen in shops and restaurants where mirrors are fixed to opposite walls.

The positions of these images and the paths of the rays can be easily determined as in the last article. Rays of light proceeding from an object O (Fig. 25) and falling on the mirror M appear, after reflection, to proceed from the primary image $I_1$. If these rays fall upon the mirror M', and are again reflected, they appear to come from a point $I_2$, which is as far behind M' as $I_1$ is in front of it. $I_2$ is a secondary image of $I_1$. In the same way, a single reflection from M' produces a primary image at $I_1'$, and a second reflection from M gives a secondary image at $I_2'$. The tertiary images $I_3$ and $I_3'$ are produced by three successive reflections (two from one mirror and one from the other). All the images lie on the straight line through O perpendicular to the mirrors. Each pair of images gives rise to another pair, for every image is in front of one or other of the mirrors. Thus in theory there should be an infinite

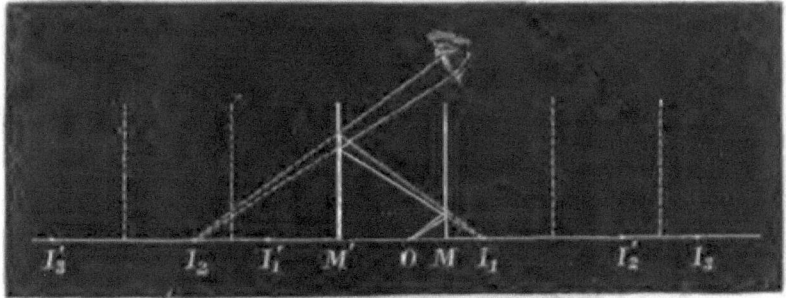

Fig. 25.

number of images: but in practice it is found that the light is so much weakened by successive reflections that beyond a certain point the images are no longer visible.

25. Images are produced not only by reflection from silvered glass but also by other reflecting surfaces, such as still water and unsilvered glass. As you walk past shop windows you see images of yourself and of other objects by reflection from the window-glass. Since the glass is transparent, you can at the same time see the contents of the shop-window. This is the principle upon which "Pepper's Ghost" and other optical illusions depend.

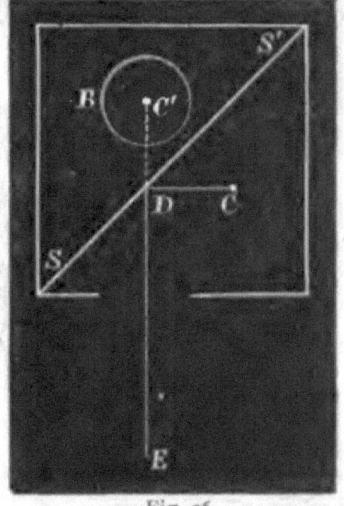

EXPT. 16.—Remove the front of a wooden box and place a beaker full of water inside it, as shown in plan at B (Fig. 26). Diagonally across the box fix a vertical sheet SS' of window-glass. At one side place a lighted candle C. Cut a hole in the front of the box and replace it. An observer looking in through this hole sees a candle apparently burning inside the beaker of water.

Fig. 26.

EXPT. 17.—Hold the point of a pencil against a sheet of thick plate-glass and look at it sideways. Several images are seen. The first of these (the one which touches the pencil point) is produced by reflection from the front surface of the glass. The second is produced by reflection

L

from the back surface. The others are produced by repeated reflections between the front and back surfaces.

Repeat the experiment with a candle, holding the glass so that the light falls on it obliquely, as in Fig. 27. If a plate-glass mirror is used

Fig. 27.

instead of a piece of unsilvered glass, the second image (that produced by reflection from the back or silvered surface) will be the brightest, as shown in the figure.

## Examples on Chapter IV

1. When a horizontal beam of light falls on a vertical plane mirror which revolves about a vertical axis in its plane, show that the reflected beam revolves at twice the rate of the mirror.

2. Distinguish between a real and virtual image formed by optical means. A candle is placed in front of a piece of flat glass, and on looking into the glass an image of the candle is seen: show how to determine the position of this image. Is it real or virtual?

3. Describe and explain the difference of the effects observed when the sun sets over a smooth lake or sea, according as the water is (1) absolutely smooth, or (2) covered with ripples.

4. A source of light is seen by reflection in two vertical plane mirrors placed against the two walls in a corner of a square room. Construct a figure showing exactly the path of a beam which enters the eye of an observer after two reflections, one at each mirror.

5. A candle is placed at a given small **distance in front of** an ordinary looking-glass made of thick plate-glass quick-silvered on the back, and a person looking obliquely into the mirror sees several images of the candle: explain this, and show the exact positions of the images by a diagram.

6. A ray of light is reflected successively from two mirrors inclined at right angles to each other (as in Fig. 24). Prove that the ray after a second reflection is parallel to its original direction.

7. Prove that if an object in front of a plane mirror moves through a distance $d$ away from the mirror, the image will move through the same distance; whereas if the mirror moves parallel to itself through a distance $d$ (the object remaining fixed) the image will move through a distance $2d$.

# CHAPTER V

## SPHERICAL MIRRORS

**26. Definitions.**—A spherical mirror is a portion of a spherical reflecting surface. The *centre of curvature* of the mirror is the centre of the sphere of which the mirror forms a part. The radius of the sphere is called the *radius of curvature* of the mirror. Spherical mirrors may be either concave or convex. In concave mirrors the reflection takes place from the inner face (that which faces *towards* the centre of curvature). In convex mirrors the reflection takes place from the outer face (that which faces *away from* the centre of curvature).

Fig. 28 represents a section of a concave spherical mirror of which $c$ is the centre of curvature. MM' is called the diameter or *aperture* of the mirror; the angle McM', included between the lines M$c$ and M'$c$, is called the *angular aperture* of the mirror. The centre of the face of the mirror is at $d$. This point is sometimes called simply the centre of the mirror. This, however, might leave us in doubt as to whether $c$ or $d$ was referred to. We shall, therefore, call the point $d$ the vertex or pole of the mirror. The distance $cd$ is the radius of curvature.

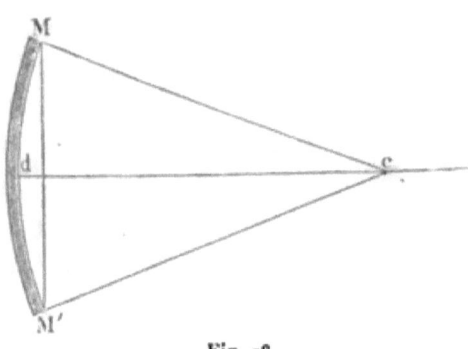

Fig. 28.

The *principal axis* of a mirror is the line passing through the vertex of the mirror and its centre of curvature. Any other line passing through the centre of curvature is called a *secondary axis*. Thus in Fig. 28 $cd$ is the principal axis, and $cM$ and $cM'$ are secondary axes.

A line drawn from any point on the mirror to its centre of curvature is a radius of the sphere, and is normal to the surface of the mirror at that point. Hence any ray of light passing through the centre of curvature and falling on the mirror is incident normally and must be reflected back along its own path. In Fig. 28, if $c$ were a luminous point sending out rays of light to the mirror, all these rays would be reflected back along their own paths and would again converge to a focus[1] at $c$.

In general, if you wish to find the direction in which a ray of light is reflected from a mirror (concave or convex), proceed as follows. Join the point of incidence to the centre of curvature of the mirror; this gives you the direction of the normal at the point of incidence. The angle between the incident ray and the normal is the angle of incidence. From the point of incidence draw a line making with the normal (but on the opposite side) an angle equal to the angle of incidence. This is the path of the reflected ray.

## Concave Mirrors

### 27. Principal Focus.

—Let us begin by considering

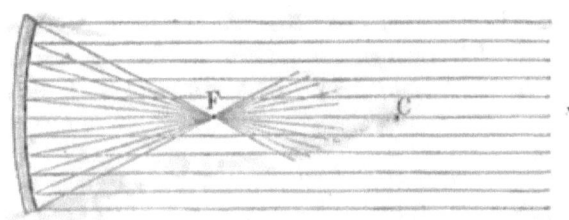

Fig. 29.—Concave Mirror and Focus.

what happens when a beam of parallel rays fall upon a concave mirror, each ray in the incident beam being parallel to the principal axis (Fig. 29). We shall prove that all the reflected

[1] A *focus* is a point towards which rays of light converge, or from which they diverge.

rays pass approximately through a single point F on the principal axis, called **the principal focus**, and that this point is midway between the **mirror and its centre of curvature.**

Let *ab* (Fig. 30) be any ray incident upon the mirror at *b* and parallel to its principal axis. C is the centre of curvature of the mirror, *d* its vertex, and C*d* the principal axis. Join *b*C. This is the normal at the point of incidence, and the angle *ab*C or *i* is the angle of incidence. From *b* draw *b*F, cutting the principal axis at F and making with the normal an angle C*b*F or *r* equal to the angle of incidence: *b*F is the reflected ray.

Now, since *ba* and *d*C are parallel and *b*C meets them, the angles *i* and *x* are equal. But $i=r$; therefore also $x=r$. Thus the triangle F*b*C is isosceles, and FC = F*b*. Now, *if we*

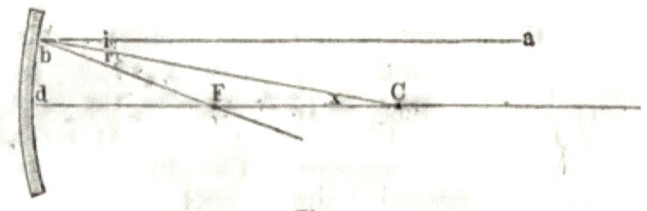

Fig. 30.

*assume* that *db* is very small compared with the radius of curvature, F*d* will be very nearly equal to F*b*. Thus FC = F*d* approximately. F is a fixed point and is midway between the vertex of the mirror and its centre of curvature.

F is called the principal **focus of** the mirror, and the distance *d*F (from the mirror **to the principal focus**) is called **the principal focal distance or focal length** of the mirror. Clearly the focal length is one-half **the radius of curvature.** If we denote the focal length by *f* and the radius of **curvature by** *r*, then $f = \frac{r}{2}$.

Any incident ray parallel to **the principal axis is** reflected so as to pass (approximately) through the **principal** focus F. Conversely, if F*b* (Fig. 30) were an incident ray, *ba* would clearly be the corresponding reflected ray. Thus any incident ray whose path **passes** through the principal focus will, after reflection, be **parallel to the** principal axis.

The assumption made in the above proof is equivalent to supposing that the angular aperture of the mirror is small, or, referring to Fig. 28, that

MM' is small compared with *dc*. We shall always assume that this is the case; and shall thus be able to obtain simple approximate relations which are sufficiently correct for most practical purposes. The student should, however, observe carefully at what stage in the reasoning the approximation comes in.

**28. Conjugate Foci.**—Let P (Fig. 31) be a luminous point on the principal axis of a concave mirror. We wish to find what becomes of the light from P after it has been reflected by the mirror.

Let PR be *any* one of the incident rays. Join R to C, the centre of curvature of the mirror. Draw the reflected ray

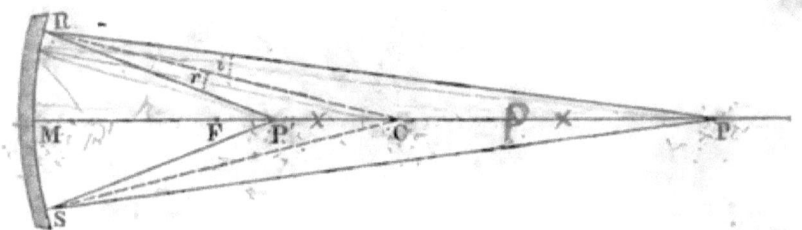

Fig. 31.—Conjugate Foci.

RP', cutting the principal axis at P', and making with the normal RC an angle $r$ equal to the angle of incidence $i$.

In the triangle RP'P the angle at R is bisected by the line RC, which also cuts the base P'P into the segments P'C and CP : hence it follows [1] that these segments are to one another in the same ratio as the sides P'R and PR of the triangle, or

$$P'C : CP = P'R : PR.$$

[*Approximation.*]—Now, if we assume, as in the last article, that the angular aperture of the mirror is very small, or that R is very near to M, we may take P'M as being very nearly equal to P'R, and PM to PR. Thus we have, approximately,

$$P'C : CP = P'M : PM \qquad . \qquad . \qquad . \quad (1)$$

Thus the distances of the points P and P' from the centre of the mirror *are in the same ratio as their distances from the mirror itself.*

---

[1] Euclid, Bk. VI. Prop. III.—'If the angle of a triangle be divided into two equal angles by a straight line which also cuts the base, the segments of the base shall have the same ratio which the sides of the triangle have to one another.'

Now let the distance PM be denoted by $p$, P'M by $p'$ and CM (the radius of curvature of the mirror) by $r$. Then

$$P'C = CM - MP' = r - p',$$

and
$$CP = PM - MC = p - r.$$

Introducing these symbols into the proportion (1), we get

$$(r-p'):(p-r) = p':p,$$

$$\text{i.e. } \frac{r-p'}{p-r} = \frac{p'}{p},$$

$$\therefore p(r-p') = p'(p-r),$$

or
$$pr - pp' = pp' - p'r.$$

Transposing, and then dividing by $pp'r$, we have

$$p'r + pr = 2pp',$$

and
$$\frac{1}{p} + \frac{1}{p'} = \frac{2}{r} \quad . \quad . \quad . \quad . \quad (2)$$

We have already seen that the focal length $f$ of the mirror is equal to $\frac{r}{2}$. Hence $\frac{2}{r} = \frac{1}{f}$, and equation (2) becomes

$$\frac{1}{p} + \frac{1}{p'} = \frac{1}{f} \quad . \quad . \quad . \quad . \quad (3)$$

**29.** This is the fundamental equation for mirrors, and, with proper conventions as to signs, is true for convex as well as concave mirrors.

The equation shows that there is a certain relation between the distances $p$ and $p'$, *i.e.* between the positions of the points P and P'. For a given mirror $f$ is a constant quantity; hence, if the value of $p$ is given, there is only one value of $p'$ that satisfies equation (3). There is only one position of P' corresponding to any given position of P.

We started by supposing that PR was *any* one of the incident rays from P : the result shows that *all* the rays from P are reflected so as to pass through a focus at the point P'. P' is the image of P in the mirror.

Conversely, if P' were a luminous point, P would be its image. For we may reverse the paths of the rays: an incident ray along P'R would give a reflected ray along RP. All rays diverging from P' would come to a focus at P.

The points P and P' are called **conjugate foci**.

*Conjugate foci with respect to a mirror are points so related that one is the image of the other in the mirror.*

In the case above considered the foci are *real*, for the rays of light actually pass through the points. We shall presently come to cases where rays of light appear to diverge from a point but do not actually pass through it: such a point is called a *virtual focus*.

Notice that parallel rays may be regarded as proceeding from a point at an infinite distance. Now, if the luminous point P is very far off, $p$ is very great and $\frac{1}{p}$ is very small. When $p$ is infinitely great $\frac{1}{p}$ is infinitely small or $= 0$. In this case equation (3) reduces to

$$\frac{1}{f} = \frac{1}{p'}, \text{ and } \therefore p' = f = \frac{r}{2}.$$

Compare this result with Art. 27.

**30. Rules for Optical Calculations.**—In order to avoid confusion as to plus and minus signs, or positive and negative focal distances, beginners will find it well to follow strictly some such rules as the following, and to use the equations for conjugate position always in the same form.

I. All distances are to be measured from the mirror (or lens); if measured towards the right hand they are to be considered positive $(+)$; if towards the left hand, negative $(-)$.

It is convenient to suppose that the luminous object, or source of light, is always placed on the right hand of the mirror: its distance will thus always be positive. This is equivalent to saying that all distances measured from the mirror on the same side as the object or source of light are to be regarded as positive; and all distances in the opposite direction as negative.

II. The equation

$$\frac{1}{f} = \frac{1}{p} + \frac{1}{p'} \qquad . \qquad . \qquad . \qquad . \qquad (3)$$

holds good for all mirrors, both concave and convex, $f$ denoting the focal length, $p$ the distance of the object, and $p'$ the distance of the image, reckoned from the mirror itself.

III. **The focal length of a concave mirror is positive.** This will be readily understood on looking at Fig. 29: for the beam of parallel light falls upon the mirror from the right hand and is reflected through the principal focus F, which is also to the right hand of the mirror. The focal length of a convex mirror is negative: this we shall consider in due course.

IV. Whenever you have to work out any optical calculation, begin by writing down the proper equation, and do not alter the signs of the letters until you come to insert their values. Now substitute in the equation the numerical values of the known quantities, *each with its proper sign*. When any two of the three quantities $f, p$, and $p'$ are given, the third can be found from the equation. If $f$ and $p$ are given, the numerical value of $p'$ tells you how far the image is from the mirror :- if $p'$ is positive, the image is formed on the same side as the object (towards the right hand of the mirror, or in front of it); but if it carries the minus sign, the image is formed on the opposite side (towards the left hand, or behind the mirror). When the distance of the object and image ($p$ and $p'$) are given, the numerical value of $f$ gives you the focal length of the mirror, and the sign of $f$ tells you whether the mirror is concave ($+$) or convex ($-$). A geometrical diagram drawn roughly to scale will often save you from making bad mistakes.

**31. Examples.**—1. Rays of light diverging from a point 3 ft. in front of a mirror converge, after reflection, to a point 1 ft. in front of the mirror. Is the mirror concave or convex, and what is its focal length?

Here $p = 3$ ft., $p' = 1$ ft., and both are positive. The focal length required is given by the equation

$$\frac{1}{f} = \frac{1}{p} + \frac{1}{p'},$$

$$= \frac{1}{3} + \frac{1}{1} = \frac{4}{3},$$

and $\therefore f = \frac{3}{4}$ ft. $= 9$ in.

The focal length is 9 in., and, since it is positive, the mirror is concave.

2. The radius of curvature of a concave mirror is 30 cm. Rays of light diverge from a point 60 cm. in front of it: to what point will they converge after reflection?

The focal length is one-half the radius of curvature, and, as the mirror is concave, it is positive. Thus $f = 15$, and $p = 60$.

Denoting the required **distance of the point from the mirror** by $p'$, we have

$$\frac{1}{15} = \frac{1}{60} + \frac{1}{p'},$$

$$\therefore \frac{1}{p'} = \frac{1}{15} - \frac{1}{60},$$

$$= \frac{4-1}{60} = \frac{3}{60} = \frac{1}{20}.$$

Thus $p' = 20$. The rays converge to a point 20 cm. in front of the mirror.

**32. Images produced by Concave Mirrors.**—Images in plane mirrors are always virtual. Images produced by concave mirrors may be either real or virtual.

In discussing the formation of images by plane mirrors we followed the paths of two rays (a normal one and any other), and found the point of intersection of the reflected rays. This was shown to be behind the mirror: the reflected rays diverged from a virtual focus. Following a similar plan here we shall be able to find—by geometrical construction and without algebraical calculation—the nature and position of the images formed by concave mirrors.

Among the many rays which proceed from any luminous point to the mirror, there are three whose directions after reflection by the mirror are easily followed.

- (1) Any ray whose path passes through the centre of curvature falls normally upon the mirror, and is therefore reflected back along its own path.
- (2) Any ray parallel to the principal axis is reflected so as to pass through the principal focus.
- (3) Any ray whose path passes through the principal focus is reflected back parallel to the principal axis.

If we wish to find the image of any luminous point, we need only follow out the directions of two of the above-mentioned incident rays and find the point of intersection of the reflected rays; this is the image of the luminous point. If the reflected rays actually intersect at this point, the image is a real one: if they have to be produced backwards (behind the mirror) before they intersect, the image is a virtual one.

**33. Real Images.**—Let AB (Fig. 32) be an object in front of a concave mirror whose focus is at F and centre of curvature at C. We shall suppose the object to be beyond C, *i.e.* at a distance greater than 2*f* from the mirror.

From A draw the ray AC*m* through C. This is reflected back along its own path. The image of A lies somewhere on this line (which is the secondary axis through A). Draw also the ray A*n* parallel to the principal axis MC. This is reflected so as to pass through F. The image of A lies somewhere along *n*F: it must therefore be situated at the point of intersection of *m*C and *n*F.

Produce *n*F to cut *m*C at *a*. *a* is the image of A. If A were

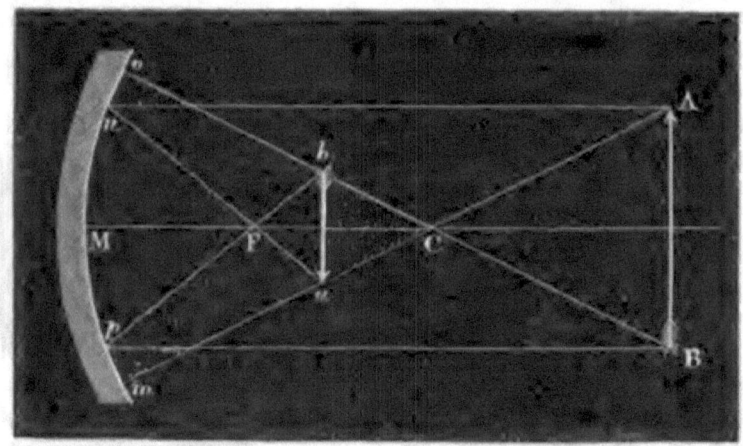

Fig. 32.—CONCAVE MIRROR: REAL IMAGE.

a luminous point, *a* would be its conjugate focus: all rays diverging from A appear, after reflection, to diverge from *a*. An eye placed in a suitable position would see a real image of A at *a*.

The image of B is found in precisely the same way by drawing the rays BC*o* and B*p*F. These intersect at *b*. *b* is the image of B. The images of points lying between A and B are formed at corresponding points between *a* and *b*. *ab* is the image of AB.

Images of this kind are produced by the bowls of spoons (a bright silver dessert-spoon is best). Hold the spoon with its concave surface towards a candle-flame and a few inches from it: looking from behind the candle towards the spoon you see a small inverted image of the flame just in front of the spoon. Hold the spoon between yourself and the light: you see a small inverted image of your face. As the bowl of the spoon is not spherical the images are, of course, distorted.

The image is *real*. Not only can it be seen by an eye placed in a suitable position, but it can be thrown upon a screen (see Expt. 19). In Fig. 32 the image is *diminished* or smaller in size than the object: this is always the case when the object is beyond the centre of curvature (C).

Observe that in all cases where a real image is formed, the image and object may change places. Thus, if *ab* (Fig. 32) were a luminous object, AB would be its image. For the paths of any of the rays may be reversed, an incident ray *aFn* giving a reflected ray *n*A, and so on. The image thus produced would be *enlarged* or magnified: this is always the case when the object is between C and F. (The student should draw the diagram for this case, using the construction given above.)

The object and its real image are formed on opposite sides of C: and, since the rays which pass through C cross each other at this point, the real image is always inverted.

We have thus arrived at the following results:—

When the object is situated beyond C (the centre of curvature), the image is formed between C and F, and is real, inverted, and diminished.

When the object is situated between C and F, the image is formed beyond C, and is real, inverted, and enlarged.

**34. Relative Sizes of Image and Object.**—In order

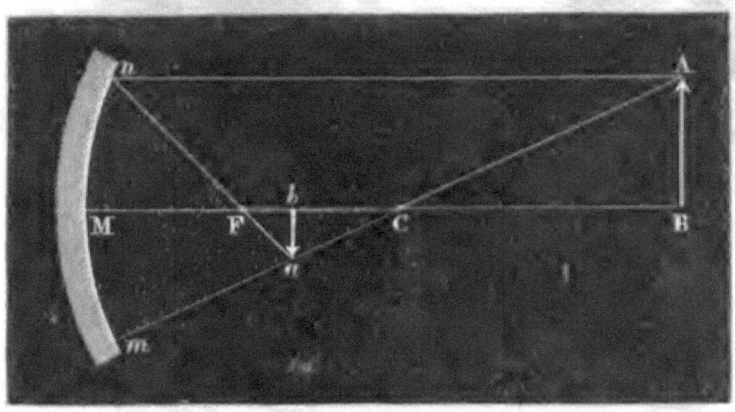

Fig. 33.—RELATIVE SIZES OF IMAGE AND OBJECT.

to simplify matters, we now take one-half of the preceding figure. AB (Fig. 33) is our object, and is perpendicular to the principal axis of the mirror. The image of A is at $a$, the position of which is found as in the last article. The image is completed by drawing $ab$ perpendicular to the principal axis (the image of B being at $b$ on this axis).

The triangles $ab$C and ABC are similar: for the opposite angles at C are equal and the angles at $b$ and B are right angles. Hence[1] the sides about these angles are proportionals, i.e.

$$\frac{ab}{AB} = \frac{bc}{BC},$$

or, *the relative sizes of image and object are as their respective distances from the centre of curvature of the mirror.*

From this it follows at once that the relative sizes are as the distances from *the mirror itself.* For $b$ is the conjugate focus of B, and in Art. 28 we saw that the distances of two conjugate points from the centre of curvature are proportional to their distances from the mirror itself. Thus

$$\frac{bc}{BC} = \frac{bM}{BM},$$

and $\therefore \dfrac{ab}{AB} = \dfrac{Mb}{MB}$.

Thus *the relative sizes of image and object are as their respective distances from the mirror.*

If we denote the sizes of image and object by I and O, and their respective distances from the mirror by $p'$ and $p$, this relation may be written in the form

$$\frac{I}{O} = \frac{p'}{p} \qquad . \qquad . \qquad . \qquad . \qquad (4)$$

and should be carefully remembered. It is true for all images, real or virtual, formed by spherical mirrors, and the student should follow out the proof for himself as we come to each special case. The ratio $\left(\dfrac{I}{O}\right)$ between the sizes of image and object is called the *magnification.* It should be noticed that

---

[1] Euclid, Bk. **VI.** Prop. **IV.**—'The sides about the equal angles of equiangular triangles are proportionals.'

the word 'size' here refers to *linear* dimensions (length or breadth) and not to area.

When the image and object are at equal distances from the mirror they are equal in size. This happens when the object is situated at the centre of curvature of the mirror (in which case $p' = p = r$).

**35. Virtual Images.**—We have yet to find what sort of an image is produced when the object is situated between the mirror and its principal focus.

Let AB (Fig. 34) be the object. From A draw the ray parallel to the principal axis and reflected from the mirror through F. Join CA and produce it to meet the mirror normally. These two reflected rays do not intersect in front of

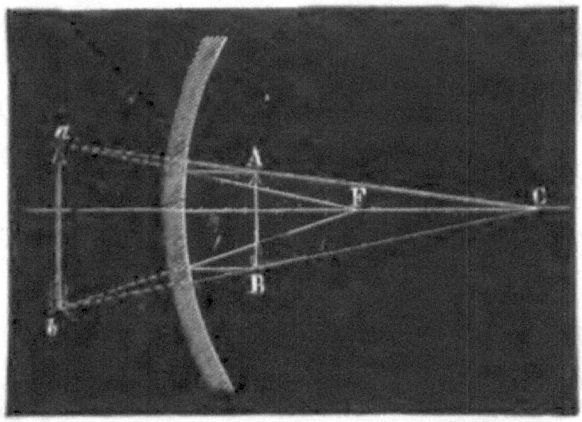

Fig. 34.—Virtual Image in Concave Mirror.

the mirror, but if produced backwards (as shown by the dotted lines) they meet at *a*. To an observer looking at the mirror all the reflected rays which originally came from A would appear to proceed from a virtual focus at *a*. *a* is the virtual image of A. In the same way *b* is the image of B. *ab* is the virtual image of AB: such an image cannot be caught on a screen, as a real image can.

Thus when an object is situated between a concave mirror and its principal focus, the image produced is *virtual, erect, and enlarged*.

**36. Experimental Illustrations.**—The student should perform the following experiments with a concave mirror, and then proceed to measure its focal length carefully by the methods described in Art. 37. The exact size and curvature of the mirror used is not of importance, but one having a focal length of about 1 foot or 30 cm. will be found convenient.

EXPT. 18.—Watch your chance when the sun is shining, and fix the mirror so that it faces the sun (the axis of the mirror pointing towards it). Move a small piece of white paper or cardboard backwards and forwards in front of the mirror until you find the position in which a sharp image of the sun is thrown on it. The image is formed at the principal focus. Measure the distance from the image to the mirror; this is the focal length.

Hold your hand so that the image is formed on the back of it; the spot is uncomfortably hot. Not only the light, but also the heat-radiation from the sun, is concentrated at this point; hence the term *focus*, which literally means a *hearth* or fireplace. If the sun is shining brightly, paper and chips of wood placed at the focus can be set on fire.

EXPT. 19—Take the mirror into a darkened room. At one end of the room place a lighted candle; at the other end place the mirror, facing the candle and at the same height as it. Find the position of the image by catching it on a small screen, as in the last experiment; if the screen cuts off too much of the candle-light, tilt the mirror so that the image is thrown slightly to one side.

The image is real, inverted, and smaller than the flame itself. Measure the distance from the mirror to the image and compare it with the distance found in the last experiment. If the candle is far enough off (20 ft. or more), the image is formed very nearly at the principal focus.

Lower the candle: the image rises. Raise the candle: the image is depressed. (See Fig. 32, which shows that the image of a point A lies on the secondary axis AC$m$.)

EXPT. 20.—Move the candle up towards the mirror. The image advances to meet it and increases in size; but remains real, inverted, and smaller than the object.

When the candle is at the centre of curvature of the mirror the image is formed at the same distance in front of the mirror (coincident with the object). In order to see this image you should tilt the mirror to one side, so that the image can be thrown on a screen placed just at the side of the candle. The image is real, inverted, and equal in size to the object.

Now move the candle nearer to the mirror. The image moves farther and farther away, and is real, inverted, and enlarged. Observe that in all the above cases the object and its real image may change places: test this by marking the positions of the candle and screen and interchanging them.

When the candle is at the principal focus the image is formed at an infinite distance. The reflected rays form a parallel beam; this is the way in which concave reflectors are used in lighthouses and " search-lights " (projectors).

When the candle is still nearer the mirror (*i.e.* between the mirror and its focus) the reflected rays diverge from virtual foci behind the mirror: no real image is formed, but on looking into the mirror you see a virtual, erect, and magnified image of the flame.

**37. Methods of Finding Focal Length.**—The focal length of a concave mirror can be measured by any of the following methods—

I. By allowing a beam of parallel rays to fall on the mirror. Rays from any very distant object, such as the sun, can be used: an image of the object is formed at the principal focus. This method has been described in Expt. 18: although it may appear the simplest, it is not the best or the most convenient.

II. Throw a real image of a candle-flame on a screen by means of the mirror. Measure the distance ($p$) from the mirror to the candle, also the distance ($p'$) from the mirror to the screen. The focal length ($f$) can then be calculated from the equation

$$\frac{1}{f} = \frac{1}{p} + \frac{1}{p'}.$$

The apparatus used (optical bench, etc.) is described on pp. 128, 129. The screen on which the image is thrown may be of white cardboard or of paper stretched on a suitable frame: it should be moved backwards and forwards until the image thrown on it is as bright and sharp as possible. This operation is called focusing, and should be carefully performed.

Make several measurements, placing the object at different distances and using both enlarged and diminished images. Calculate the value of $f$ corresponding to each pair of values of $p$ and $p'$ obtained, and take the mean of the results.

III. When an object is placed at the centre of curvature of a mirror, the image coincides in position with it. Thus if the object is a luminous point at $c$ (Fig. 35), all rays diverging from it fall normally on the mirror, and are, therefore, reflected back along their own paths to $c$. The image is a luminous point coincident with the object. The distance $dc$ is the radius of curvature of the mirror, and is therefore twice the focal length. This method of finding the focal length of a mirror is simple and accurate, and can be carried out as follows :—

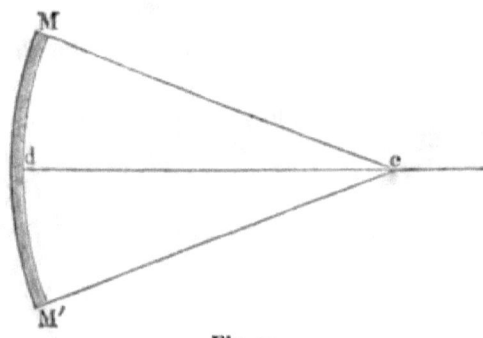

Fig. 35.

Take a piece of cardboard about 4 in. square, and in the centre of it cut out with a cork-borer a hole about ¼ in. in diameter: across the hole stretch two hairs or black threads at right angles. This is to be our object or source of light. Mount the cardboard screen on one of the uprights of the optical bench so that the hole is at the same height as the centre of the mirror. Illuminate the hole and cross-threads strongly by placing an Argand burner behind (*i.e.* on the side farther from the mirror). The light proceeding from the hole is reflected by the mirror to a real focus (provided the mirror is not too near). This focus may be in front of the screen or behind it. You have now to find *by trial* a position of the mirror such that this focus is on the screen itself.

You see on the screen a blurred spot of light. Adjust the

mirror so that this spot falls just by the side of the hole. Move the mirror backwards or forwards until the spot of light becomes a sharp bright image of the hole, with the cross-threads well defined. Measure the distance from the mirror to the screen: one-half of this is the focal length.

**38. Table of Results.**—After performing the above experiments the student should find for himself the position of the image for any given position of the object, using both the algebraical equations and geometrical diagrams. He should also find the nature of the image produced and again verify his results by experiment.

In the following table C denotes the centre of curvature of the mirror, F its principal focus, and M the mirror itself.

| Position of Object. | Position of Image. | Characteristics of Image. |
|---|---|---|
| At infinity | At F | Real, inverted, diminished. |
| Beyond C | Between F and C | Real, inverted, diminished. |
| At C | At C | Real, inverted, equal in size. |
| Between C and F | Beyond C | Real, inverted, enlarged. |
| Between F and M | Behind M | Virtual, erect, enlarged. |

**39. Examples.**—1. A candle-flame 1 in. long is 36 cm. in front of a concave mirror whose focal length is 30 cm. Find the nature, position, and size of the image.

Here $p = 36$, $f = 30$; and both are positive. The value of $p'$ (the distance of the image from the mirror) is given by the equation

$$\frac{1}{f} = \frac{1}{p} + \frac{1}{p'}.$$

Substituting the above values we have

$$\frac{1}{30} = \frac{1}{36} + \frac{1}{p'}.$$

or
$$\frac{1}{p'} = \frac{1}{30} - \frac{1}{36}$$
$$= \frac{6-5}{180} = \frac{1}{180},$$

and $\therefore p' = 180$.

The distance of the image from the mirror is 180 cm. Since $p'$ is positive, the image is formed in front of the mirror, and is real and inverted. (Verify this by sketching the diagram.)
The relative sizes of image and object are given by the equation
$$\frac{I}{O} = \frac{p'}{p} \quad \text{(see p. 158).}$$

Hence
$$\frac{I}{O} = \frac{5}{1} = 5.$$

The image is five times as large as the object, and is 5 in. long.

2. *The candle-flame is placed at a distance of* 15 *cm. from the same mirror. What sort of an image is now produced, and what is its size?*

As before, we have
$$\frac{1}{30} = \frac{1}{15} + \frac{1}{p''}$$

and
$$\therefore \frac{1}{p'} = \frac{1}{30} - \frac{1}{15} = -\frac{1}{30}.$$

Hence $p' = -30$ cm. This means that the image is formed 30 cm. *behind* the mirror, and is virtual and erect.
The distances of the image and object from the mirror are as 30 to 15, or as 2 to 1. The image is, therefore, twice the size of the object, or is **2** in. long.

3. *You are required to throw upon a wall an image of a gas-flame which stands 8 ft. from the wall, and the image is to be three times the size of the flame. What sort of a mirror would you choose, and where would you hold it?*

Suppose the mirror to be placed $x$ feet from the object on the side farther from the wall; it will then be $(8+x)$ ft. from the wall on which the image is to be thrown. Our distances now are
$$p = x, \quad \text{and } p' = 8 + x.$$

Further, since the image is to be three times the size of the object, we must have $p' = 3p$, or
$$(8 + x) = 3 \times x, \quad \text{and } \therefore x = 4.$$

Thus $p = 4$, $p' = 8 + 4 = 12$,

and
$$\frac{1}{f} = \frac{1}{4} + \frac{1}{12} = \frac{1}{3},$$

$f = +3$. This means that the mirror required is a *concave* mirror of 3 ft. focal length. We have already found that $x = 4$; thus the mirror must be held 4 ft. from the gas-flame, or 12 ft. from the wall.

[Observe that the distances of image and object from the mirror bear the right ratio to one another, being as 12 to 4 or 3 to 1. Check your results in this way whenever you can.]

## Convex Mirrors.

**40. Focal Length Negative.**—Suppose a beam of parallel rays to fall upon a convex mirror, each ray in the incident beam being parallel to the principal axis of the mirror (CM, Fig. 36). Let PR be any one of the

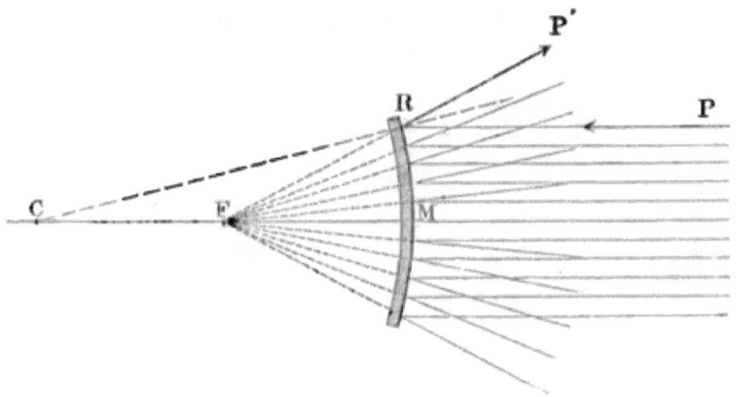

Fig. 36.—Convex Mirror and Principal Focus.

incident rays. Draw the normal CR and produce it. The ray PR is reflected along RP', making with the normal an angle equal to the angle of incidence.

Produce P'R backwards (behind the mirror, as shown by the dotted line) to cut the axis at F. As in the case of concave mirrors, Art. 27, it can be proved that F is (approximately) midway between C and M. F is the principal focus of the mirror. All incident rays parallel to the principal axis appear after reflection to *diverge* from F.

MF is the *focal length* of the mirror and is one-half the radius of curvature (CM). Observe that F lies *behind* the mirror. The incident light comes from the right hand, the principal focus is on the *left hand* of the mirror. Hence, according to our rules for signs (Art. 30, Rule I.), *the focal length of a convex mirror is negative*.

**41. Virtual Images.**—An incident beam of parallel rays falling upon a convex mirror is converted into a *divergent* beam. If the incident beam is already divergent, the divergence is *increased* by reflection from the mirror. Hence convex mirrors cannot give rise to real images.

Let AB (Fig. 37) be an object in front of a convex mirror. Through A draw A*n* parallel to the principal axis MC: the incident ray A*n* is reflected along *nr* so that it appears to come from the focus F. Join AC: the incident ray A*m* is reflected back along its own path. The two reflected rays do not intersect in front of the mirror, but the dotted lines F*n* and C*m* intersect at *a*. To an observer looking at the mirror all rays proceeding from A appear, after reflection, to proceed from *a*. *a* is the virtual image of A.

Similarly *b* is the image of B. *ab* is the virtual image of the object AB. The image is *erect*; for image and object are on the same side of C, and hence the lines AC and BC do not cross each other between the object and image (see Art. 33). The image is *smaller* than the object: for their

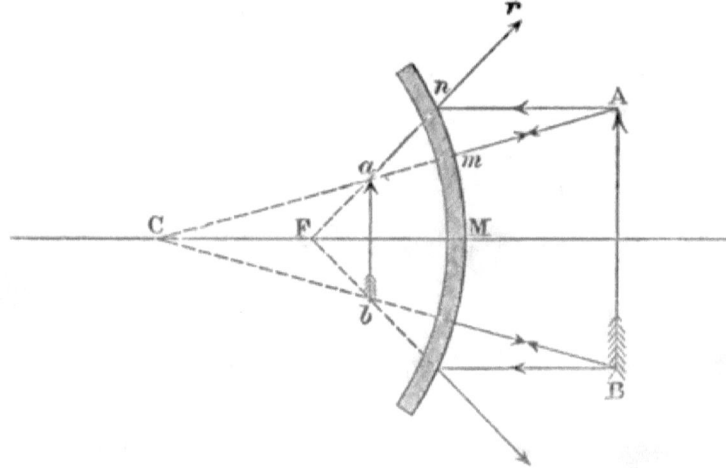

Fig. 37.—Virtual Image in Convex Mirror.

relative sizes are as their respective distances from C, and the image always lies between C and the object.

Thus images produced by convex mirrors are always *virtual, erect, and diminished*.

Such images can be seen in the convex surfaces of spoons, tea-pots, the silvered glass balls used as ornaments, etc. A fairly good convex mirror can be made by blackening or silvering the concave surfaces of a watch-glass or clock-glass.

*Example.*—An object 3 inches in length is held 6 inches in front of a convex mirror whose radius of curvature is 2 feet. Find the position and size of the image.

The focal length is one-half the radius of curvature, and is therefore 1 foot or 12 inches. But, since the mirror is convex, its focal length is negative. Thus

$$f = -12, \text{ and } p = +6.$$

The distance of the image from the mirror is given by the equation

$$-\frac{1}{12} = \frac{1}{6} + \frac{1}{p''}$$

or

$$\frac{1}{p'} = -\frac{1}{12} - \frac{1}{6} = -\frac{3}{12},$$

and

$$p' = -4.$$

Thus the image is virtual **and** is formed 4 inches **behind** the mirror.

The relative sizes of image and object are as their distances from the mirror, *i.e.* as 4 to 6, or **2 to** 3. The image is therefore 2 inches long.

### Examples on Chapter V

Read carefully the rules given in Art. 30 and observe how these are applied in the solved examples in Arts. 31-41. If your results do not agree with the answers, see whether you have given the wrong sign ( + or − ) to any of the quantities in the equations. Check your results by drawing diagrams roughly to scale.

1. What is meant by an image of an object? Why are some images called real, and others virtual? Explain and illustrate by a sketch the formation of an image of each kind by a concave mirror.

2. How would you render visible an image of the sun formed by a concave spherical mirror? Is the image real or virtual?

3. The radius of curvature of a concave mirror is 2 feet : what is its focal length? What would be the position of the image of a point 4 feet in front of this mirror?

4. Find the position of the image of a point 4 inches in front of the same mirror.

5. An object is placed 120 cm. in front of a mirror, and a real image is formed 40 cm. from the mirror and on the same side of it. What is the focal length of the mirror, and is it concave or convex?

6. A candle-flame 1 inch long is 18 inches in front of a concave mirror whose focal length is 15 inches. Find the position and size of the image.

7. Rays from a luminous point 20 cm. in front of a mirror are found, after reflection, to converge to a point 30 cm. in front of the mirror. Find the nature and focal length of the mirror.

8. An object 1 inch long is placed at a distance of 1 foot from a concave mirror of 9 inches focal length. Where is the image formed and what is its size?

9. A bright object, 4 inches high, is placed on the principal axis of a concave spherical mirror, at a distance of 15 inches from the mirror. Determine the position and size of its image, the focal length of the mirror being 6 inches.

10. A gas-jet is placed on the principal axis of a concave mirror and 1 foot in front of it. A real and inverted image is produced on a screen held in front of the mirror, but at a greater distance than the gas-jet. If the image is twice as long as the flame, what is the focal length of the mirror?

11. A real image produced by a concave mirror is found to be twice the size of the object : if the focal length of the mirror is 1 foot, where are the object and image situated ? What would be the relation between their sizes if their positions were reversed ?

12. An object 6 cm. in length is placed at a distance of 50 cm. in front of a concave mirror of 20 cm. focal length. Find the nature, position, and size of the image.

13. You wish to throw upon a wall a real image of a gas-flame, 2 feet distant from it, by means of a mirror. What kind of mirror would you use, what should be its focal length, and where would you place it?

14. An object is placed 5 inches from a concave mirror of 6 inches focal length : where is the image produced, and what is the magnification?

15. A concave mirror of 2 feet focal length is placed 1 foot from an object : find the change in the position of the image produced by moving the object 1 inch nearer the mirror.

16. Prove that the focal length of a *convex* spherical mirror is equal to half its radius of curvature. (Apply the method of Art. 27 to Fig. 36.)

17. An object 3 inches long is held 6 inches in front of a convex mirror whose radius of curvature is 2 feet. Find the nature, position, and magnitude of the image.

Where will the image be produced when the object is held 2 feet in front of the same mirror, and what will be its size?

18. An object 4 cm. long is placed at a distance of 10 cm. from a convex mirror of 30 cm. focal length. Find the position and size of the image.

# CHAPTER VI

## REFRACTION OF LIGHT

**42. Simple Experiments on Refraction.**—We have hitherto considered only the behaviour of light as it travels through a homogeneous medium (Arts. 2 and 4), in which case it is propagated *in straight lines.* But when a ray of light passes from one transparent medium into another, its direction

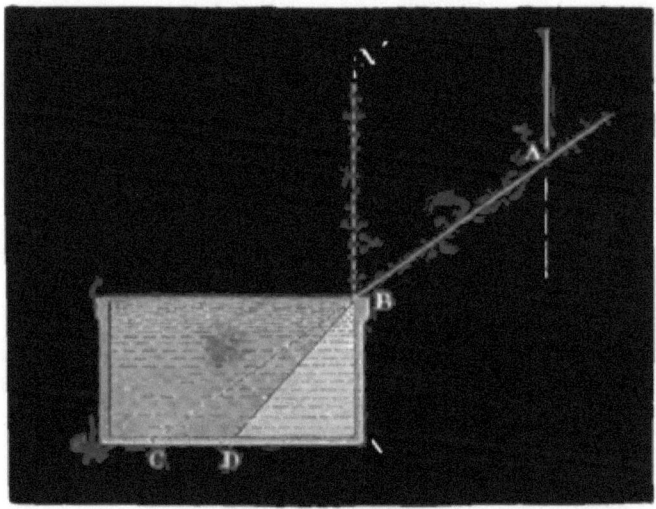

Fig. 38.—Refraction of Beam of Light.

is changed and it is said to be *refracted.* This bending or *refraction* of light may be illustrated by the following experiments.

Expt. 21.—Procure a rectangular tin box (a biscuit-box). Mark a scale of inches or centimetres along the bottom of it, or lay a metal scale on the bottom. Take the box into a dark room and allow a beam of parallel light to fall slantwise against the edge. (It is best to use a strong source of light—sun-light or lime-light.) The side of the box throws a shadow C (Fig. 38), which is in a line with the direction of the incident beam AB. Note the point C where the edge of this shadow falls on the bottom.

Now (without altering the position of anything) fill the box with water. The edge of the shadow moves to D, nearer the vertical side BN of the box. Clearly the light is refracted or bent on entering the water. Note the *direction* in which it is refracted. N'N is the normal at the point of incidence. *On passing from air into water light is refracted towards the normal.*

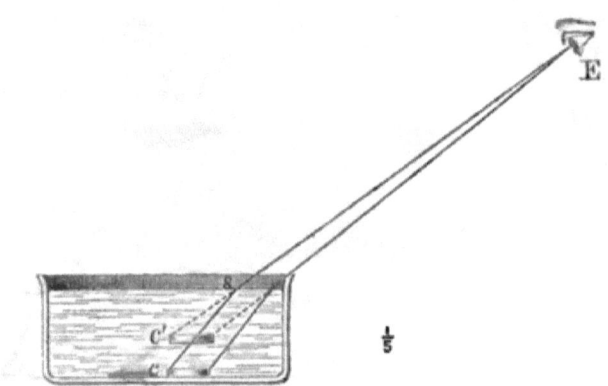

Fig. 39.—APPARENT DISPLACEMENT OF COIN.

EXPT. 22.—Put a coin on the bottom of an empty basin (*c*, Fig. 39). Place your eye in such a position (E) that you just cannot see it—the coin being hidden by the side of the basin.

If water is now poured into the basin (the eye still remaining at E), the coin becomes visible and appears to rise as the level of the water rises. This would clearly be impossible without some bending of the rays of light which proceed from the coin. A ray such as *cs* (which would otherwise not reach your eye) is refracted downwards along *s*E on leaving the water and thus enters your eye. The eye takes no notice of the refraction: it simply sees the coin at *c'* along E*s* produced.

Notice the *direction* in which the light is refracted. *On passing from water into air light is refracted away from the normal.*

EXPT. 23.—Plunge a stick or pencil slantwise into water. It looks as though it were bent just at the surface of the water, and the part immersed appears shortened and elevated (Fig. 40). This is best seen by placing the eye at one side: if you look *along* the stick, or if it is held upright in the water, it simply appears shortened and not bent.

Fig. 40.—APPEARANCE OF STICK IN WATER.

**43. Laws of Refraction.**—Let RI (Fig. 41) represent a ray of light in air incident obliquely at I upon the surface of

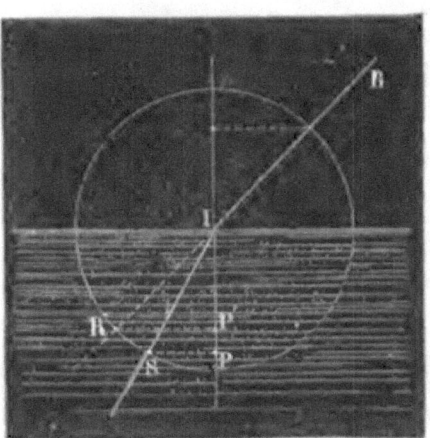

Fig. 41.—REFRACTED RAY.

another medium, *e.g.* still water. Draw the normal at I to the refracting surface. The angle between the normal and the incident ray RI is called the *angle of incidence*. We know that on entering the water the ray of light is bent in some such direction as IS—*i.e.* towards the normal. The angle SIP between the normal and the refracted ray IS is called the *angle of refraction*. Clearly the angle of refraction is less than the angle of incidence.

Produce RI to R'. The angle R'IS measures the amount by which the ray is bent or *deviated* out of its path: it is called the *angle of deviation*. Whenever a ray of light in passing from one medium to another is bent towards the normal, the second medium is said to be 'optically denser' than the first. Thus water and glass are both optically denser[1] than air.

With I as centre describe a circle. From the point where this circle cuts the incident ray draw a perpendicular to the normal: this is clearly equal in length to the perpendicular R'P' drawn from the point R' where the circle cuts the incident ray produced. Again, from the point where the circle cuts the refracted ray draw a perpendicular to the normal. The *ratio* of these perpendiculars has *a constant value* for any given pair of media (air and water, air and glass, etc.)

We may now state the laws of refraction in the following form :—

I. *The refracted ray lies in the plane containing the incident ray and the normal, and on the opposite side of the normal.*

II. *If points equidistant from the point of incidence be taken on the incident and refracted rays, and if from these points perpendiculars be drawn to the normal, the ratio of these perpendiculars is constant for any given pair of media.*

44. This constant ratio is called the index of refraction for the two media, and is usually denoted by the letter $\mu$. Its value for air and water is about $\frac{4}{3}$: for air and glass about $\frac{3}{2}$; but it may be greater than this according to the composition of the glass.

The following are approximate values of the indices of

[1] This use of the term 'dense' must not be confused with the usual meaning of density as defined on pp. 24, 25. There is no necessary connection between the two. Thus turpentine is lighter than water but is optically denser: it floats upon the surface of water, but a ray of light in passing from water into turpentine is deviated towards the normal.

refraction for light passing from air into certain transparent media.

| | |
|---|---|
| Water | 1·33 |
| Alcohol | 1·37 |
| Turpentine | 1·48 |
| Benzene | 1·50 |
| Carbon Bisulphide | 1·67 |
| Common Window-Glass (about) | 1·53 |
| Crown Glass (Potash-lime-glass free from lead) | 1·53 to 1·61 |
| Flint-Glass (Lead-glass) | 1·62 to 1·79 |
| Diamond | 2·47 |

**45. Experimental Verification.**—The first law of re-

Fig. 42.—Verification of Law of Refraction.

fraction offers no difficulty: the second can be verified by means of the apparatus shown in Fig. 42. This consists of a shallow cylindrical glass vessel, the lower half of which is filled with water. The vessel is surrounded by a divided circle on which the angles of incidence and refraction can be measured. The perpendicular distances corresponding to SP and R'P' in Fig. 41 can also be measured on the horizontal scale which slides up and down on a vertical rod. By this means the second law of refraction can be verified directly.

EXPT. 24.—In order to show up the path of the beam as it passes through the water, a few drops of milk must be added beforehand (just enough to make the water milky). Pour in the water through the hole at the top of the glass vessel until it rises just to the centre-mark O. Blow in a little smoke through the hole so as to show up the path of the beam in the upper half of the vessel.

Direct the beam of light on the movable mirror I and adjust this so that the light is reflected through the slit in the tube (in front of I) and on the surface of the water at O. The movable arm O$a$ which carries the mirror is now parallel to the incident beam.

The beam of light is refracted along OR and emerges at R without undergoing further refraction, for the beam is normal to the surface of the vessel at R. Adjust the movable arm $b$ so that the emergent beam passes through the slit in the tube attached to it: O$b$ is now parallel to the emergent beam. The angle $a$O$b$ measures the deviation.

Measure off the distance from the centre of the horizontal scale to the point $b$: this distance corresponds to the length of the perpendicular SP in Fig. 41. Raise the scale and in the same way measure off the distance O'$a$: this corresponds to the length of the perpendicular R'P' in Fig. 41.

Alter the position of the arm O$a$ and repeat the adjustments and measurements. Whatever the angle of incidence, it will be found that *the ratio of the perpendiculars is constant.* Thus if O$a$ is 6 cm., O$b$ will be 4·5 cm.: if O$a$ is 8 cm., O$b$ will be 6 cm., *the ratio being always as 4 to 3.*

Place the arm O$a$ vertical and reflect the light from the mirror vertically downwards: the light is now incident normally on the surface of the water and *passes straight through without any deviation.* Tilt the arm gradually and measure the deviation for each angle of incidence: *the deviation increases with the angle of incidence.*

The light is not *all* refracted: some of it is reflected from the surface of the water according to the usual law. Notice that this reflected beam becomes brighter as the angle of incidence increases. The amount of light reflected from the surface of water *increases with the angle of incidence.*

Further experiments may be made with another liquid, *e.g.* petroleum. Its index of refraction will be found to be constant, but different to that of water.

It will generally be found sufficient to illustrate the law without verifying it by measurements, and in this case a simple glass tank will do well instead of the apparatus above described. A convenient form of tank for refraction experiments can be made out of a tin biscuit-box. Leave the top of the box open, cut a slit (about 1 inch broad) down one of the sides, and fasten a strip of glass water-tight against this with marine glue. Cut a circle out of the front of the box (or cut the front out leaving a rim round) and fasten a sheet of glass in the same way against it. A beam of light can be admitted either from the top or the side through a slit in a blackened card. Even a common medicine-bottle with flat sides can be used: it should be painted over with black paint, one side and a circle on the front being left clear (see Fig. 45).

Instead of adding milk to the water, a little common salt may be dissolved in it and a few drops of a very dilute solution of silver nitrate added. The slight precipitate of silver chloride thus produced will remain for some time in suspension.

### 46. Geometrical Construction for Refracted Ray.—

We shall first consider the case of a ray passing *from a rarer to a denser medium*, *e.g.* from air into water or glass ($\mu > 1$). Suppose the paths of the incident and refracted rays to be as in Fig. 43. At the point of incidence C draw the normal to the refracting surface AB. With C as centre draw a circle cutting the incident and refracted rays at I and R respectively.

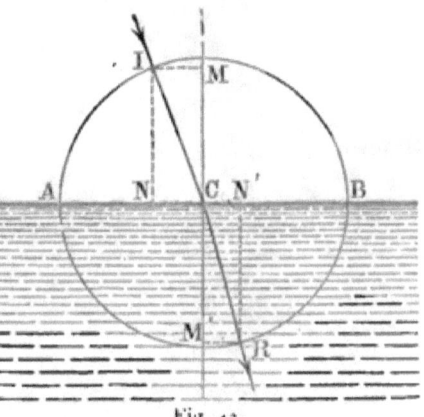

Fig. 43.

From I draw IM and from R draw RM′ perpendicular to the normal. The ratio of these two perpendiculars is, by definition

(Art. 44), the index of refraction. Thus if the two media are air and water,

$$\frac{IM}{RM'} = \text{(air)}\ \mu\ \text{(water)} = \frac{4}{3}.$$

Draw IN and RN' perpendicular to the surface AB. Clearly $IM = CN$ and $RM' = CN'$.

Thus $\quad\dfrac{CN}{CN'} = \dfrac{4}{3},\quad$ or $CN' = \dfrac{3}{4} \cdot CN$.

Hence the following geometrical construction for finding the path of the refracted ray.

*Given the path of a ray in air, to find the path of the refracted ray in water* $\left(\mu = \frac{4}{3}\right)$. Let IC be the incident ray. From I draw IN perpendicular to the surface AB. Divide CN into *four* equal parts. From C, on the opposite side, lay off CN' equal to *three* of these parts. At N' draw a perpendicular to the surface AB. With C as centre and CI as radius describe a circle cutting this perpendicular at R. Join CR. CR is the path of the refracted ray.

If the second medium be glass $\left(\mu = \frac{3}{2}\right)$, modify the construction as follows. Divide CN into *three* equal parts. From C, on the opposite side, lay off CN' equal to *two* of these parts.

In general, if the index of refraction be $\mu$, CN' must be taken equal to $\dfrac{1}{\mu} \cdot CN$.

Notice that as $\mu > 1$, $CN' < CN$. Hence the point N' must fall inside the circle (to the left of B), for, however great the angle of incidence, CN cannot be greater than CA (the radius). Hence also the perpendicular from N' will always cut the circle. For every possible position of the incident ray there is a corresponding refracted ray. Thus the construction is always possible when the ray is passing from a rarer to a denser medium.

**47. Refraction from a Denser into a Rarer Medium.**—In any case of refraction the incident and refracted rays may be supposed to change places (the direction in which the light travels being reversed). Thus in Fig. 43, if RC be supposed to represent an incident ray, CI would be the corresponding refracted ray in air. This is equivalent to stating that when **the direction of the ray is reversed the index of refraction is reversed.** Thus, if the index of refraction for light passing from air into a denser medium is $\mu$, the index of refraction for light passing from this denser medium into air will be the reciprocal of $\mu$, or $\frac{1}{\mu}$.

In the cases of water and glass we have

$$\text{(air)}\ \mu\ \text{(water)} = \frac{4}{3},\quad \text{(water)}\ \mu\ \text{(air)} = \frac{3}{4},$$

$$\text{(air)}\ \mu\ \text{(glass)} = \frac{3}{2},\quad \text{(glass)}\ \mu\ \text{(air)} = \frac{2}{3}.$$

*Given the path of a ray in water, to find the path of the refracted ray in air* $\left(\mu = \frac{3}{4}\right)$. Let IC (Fig. 44) be the incident ray. From I draw IN perpendicular to the refracting surface AB. Divide CN into *three* equal parts. From C, on the opposite side, lay off CN' equal to *four* of these parts. From N' draw a perpendicular to the surface AB. With C as centre and CI as radius describe a circle cutting this perpendicular at R. Join CR. CR is the path of the refracted ray.

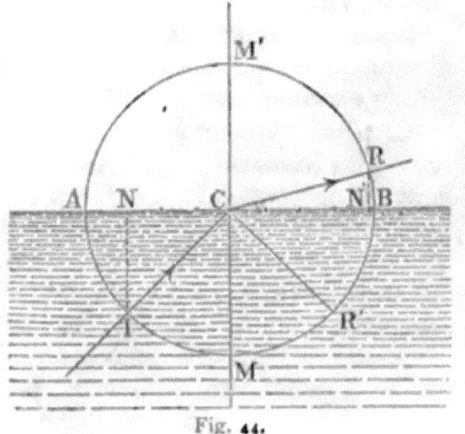

Fig. 44.

If the ray be travelling from glass into air $\left(\text{in which case } \mu = \frac{2}{3}\right)$, modify the construction as follows. Divide CN into *two* equal parts. From C, on the opposite side, lay off CN' equal to *three* of these parts.

In general, if the index of refraction be $\mu$ (reckoned in the direction in which the light is travelling), CN' must be taken equal to $\frac{1}{\mu}$. CN.

Now notice that as $\mu$ is here $< 1$, CN' > CN. Hence the point N' *may* fall outside the circle. When it does, the perpendicular through N' to the surface will not cut the circle at all.

**48. Experimental Verification.**—Before proceeding further with the question of refraction into a rarer medium, the statements already made should be verified.

**EXPT. 25.**—Using the same apparatus as in Expt. 24, throw the beam of light upward through the water, adjusting the mirror so that the beam meets the surface of the water at O. Measure off the perpendicular distances to the normal and see whether they are in the ratio of 3 to 4 (they were in the inverse ratio—4 to 3—when the ray passed from air through water), or mark the positions of the incident and refracted rays and then reverse the direction—sending the incident ray along the path previously followed by the refracted ray. Notice that the refracted ray in air now follows the path of what was before the incident ray.

Send the light vertically upwards through the water: it is now incident normally on the surface and passes straight through without any deviation.

Gradually increase the angle of incidence. The angle of refraction increases at the same time, but more rapidly. The light is not *all* refracted: some of it is reflected from the surface of the water according to the usual law, and the amount thus reflected increases with the angle of incidence.

As you increase the angle of incidence the angle of refraction increases from 0° to 90°, beyond which it clearly cannot increase. When the angle of refraction is 90° the refracted ray just grazes the surface of the water: the corresponding angle of incidence is called the *critical angle*. Its value for water is about $48°\tfrac{1}{2}$.

Just at this point a very remarkable thing happens. As soon as you increase the angle of incidence beyond the critical angle the refracted ray disappears entirely and the reflected ray increases greatly in brightness. None of the light escapes into the air: it is *all* reflected (Fig. 45). The reflection is said to be *total* or complete. It takes place according to the usual law, but differs from ordinary reflection in the respect that it is more complete, none of the light being scattered (as is the case with even the best polished mirrors).

We shall see presently that the critical angle has different values in the case of different substances, depending upon the

Fig. 45.—TOTAL REFLECTION.

index of refraction. For the present remember that our experiment has proved the following result : *If a ray of light in water is incident upon the surface at an angle greater than the critical angle, it cannot get out into the air but is totally reflected back into the water.*

**49. The Critical Angle.**—Look at Fig. 44 again. The angle of refraction (in air) is not far from 90°. If you were to increase the angle of incidence gradually, you would reach a point at which the angle of refraction would be exactly 90°, and the refracted ray would then just graze the surface of the water. For in Fig. 44 N' is not far from B. As CN increases CN' increases in the same ratio and ultimately becomes equal to CB (the radius of the circle). N' now coincides with B, and the perpendicular to the surface through N' becomes the tangent touching the circle at B (see Fig. 46). CB is the refracted ray. The corresponding angle of incidence ($\theta = $ ICM) is the **critical angle.** For any ray incident at an angle greater than this, the point N' would fall *outside* the circle. The perpendicular to the surface through N' would not cut the circle at all, and the usual construction for finding the refracted ray in air would cease to be possible. There would be *no* refracted ray : the light would suffer *total reflection.*

To find the critical angle for light travelling from water into air $(\mu = \frac{3}{4})$ proceed as follows. Take as centre any point C (Fig. 46) in the line separating the air from the water: with any radius describe a circle cutting this line at B. Divide CB into *four* equal parts. From C, on the opposite side, lay off CN equal to *three* of these parts. From N draw a perpendicular to the surface cutting the circle at I. Join IC. The angle ICM is the critical angle. If you measure it with a protractor you will find that it is about $48°\frac{1}{2}$.

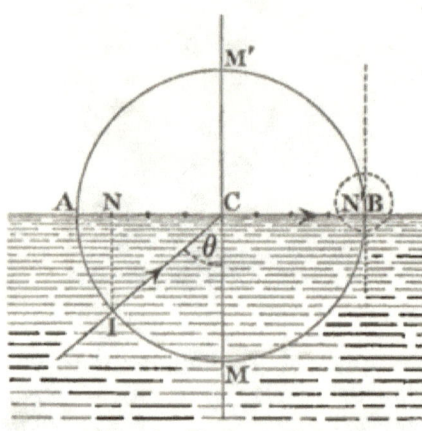

Fig. 46.—Critical Angle.

You can easily see how the construction would have to be modified for any other media. In the case of light travelling from glass to air $(\mu = \frac{2}{3})$, take CN equal to two-thirds of the radius CB. In general, if $\mu$ be the relative index of refraction, take $CN = \mu \cdot CB$.

We have hitherto supposed the refractive index to be measured *in the direction in which the light is travelling*, $\mu$ being greater or less than unity according as the light is passing into a denser or a rarer medium. In future we shall, in general, use $\mu$ to denote the relative index for refraction *from the rarer into the denser medium* $(\mu > 1)$, and when the contrary is not stated it may be assumed that the first medium is air.

Using this notation we should have to take $CN = \dfrac{CB}{\mu}$ in the above construction for finding the critical angle. Observe that as the refractive index increases the critical angle diminishes:—

| | Refractive Index. | Critical Angle. |
|---|---|---|
| Water | 1·33 | $48°\frac{1}{2}$ |
| Benzene | 1·50 | $42°$ |
| Crown Glass (about) | 1·56 | $41°$ |
| Flint Glass (about) | 1·66 | $37°$ |
| Carbon Bisulphide | 1·67 | $36°\frac{1}{2}$ |
| Diamond | 2·47 | $24°$ |

**50. Total Reflection.**—Suppose A (Fig. 47) to be a source of light under the surface of water. A ray proceeding vertically upwards from it passes out into the air without suffering any deviation. Rays on each side of this are deviated more and more as the angle of incidence increases. Beyond a certain point the rays can no longer escape into the air but are totally reflected back into the water. The critical angle separates the rays which are refracted into the air from those which are totally reflected.

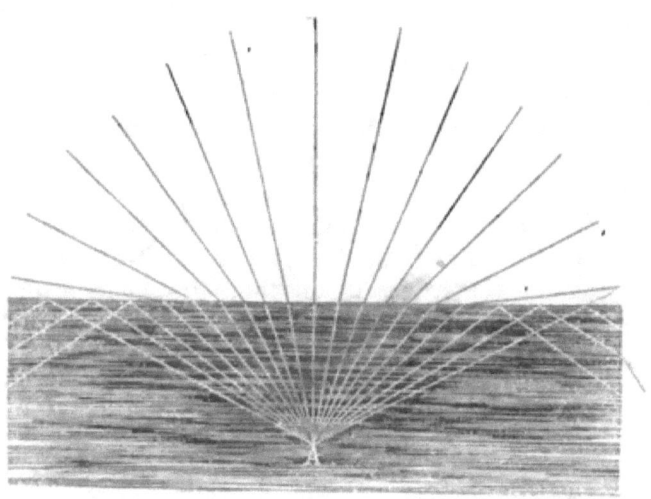

Fig. 47.

Again, suppose the direction of the rays to be reversed. All possible refracted rays reaching the point A are included within a cone of semi-vertical angle $= 48°\frac{1}{2}$ (see last article). Thus, to an observer under the water in a pond, all objects above the surface of the water (however far on either side) would appear crowded together within such a cone. On looking beyond this cone (*i.e.* in a direction more inclined to the vertical) he would only see, by total reflection from the surface of the water, objects lying farther off at the bottom of the pond.

EXPT. 26.—Hold a glass of water between your eye and the light, and raise it so that the surface of the water is above the level of your eye. Looked at thus from below, the surface of the water appears like quicksilver.

Place a spoon in the glass. You can see nothing through the surface of the water, but the lower part of the spoon can be seen by reflection more perfectly than in a mirror.

EXPT. 27.—Lay a glass prism (Art. 57) face downwards on a printed page. If you look at it directly from above, you can read the print through the glass: but if you shift your eye to one side and lower it gradually, you will come to a point at which the print suddenly becomes invisible, and the face of the prism appears brighter than a sheet of polished silver.

The same appearance is well shown by glass cubes such as paper-weights and ink-pots. Whatever may be the direction of a ray incident upon one of the sides of the cube, it is refracted so much as to make with the base an angle less than the critical angle. Consequently all the light incident upon the base of the cube is reflected with the brilliancy characteristic of total reflection.

EXPT. 28.—Place a lighted candle in such a position that it can be seen by total reflection from the internal surface of a glass prism. Then turn the prism so that the candle is seen by ordinary reflection from one of the faces. Compare the brilliancy of the two images.

EXPT. 29.—Hold an empty test-tube slantwise in water and look at it from above: it looks as if it were filled with mercury. The light which passes through the water cannot escape into the air inside the test-tube, but is totally reflected upwards.

Pour water gradually into the tube: the brilliant reflection disappears as the water rises.

EXPT. 30.—The preceding experiment can be modified so as to afford a direct comparison between the brilliancy of total reflection and that of ordinary reflection from the surface of mercury. For if the lower part of the tube be filled with mercury it will be found that the reflection from this is not nearly as bright as that from the upper (empty) half.

The tube should be inclined at about the angle shown in Fig. 48, and a sheet of white paper should be laid on the table in front (PP').

Good examples of total reflection are often shown by cracks in panes of window-glass when looked at obliquely. In many physical instruments the principle of total reflection is usefully employed: for example, in the reflecting prisms of spectroscopes. If it is desired to reflect through the slit of a spectroscope light proceeding from a source A at one side (Fig. 49), there is placed in front of the slit a glass prism, the section of which is a right-angled isosceles triangle. The light from the source A enters normally through one of the faces of the prism and falls upon the hypotenuse at an angle of incidence of 45°. Since this is greater than the critical angle (Art. 49) of the glass, the light is totally reflected. At the same time it is deviated through an angle of 90°, so that it emerges normally from the other face of the prism and enters the slit.

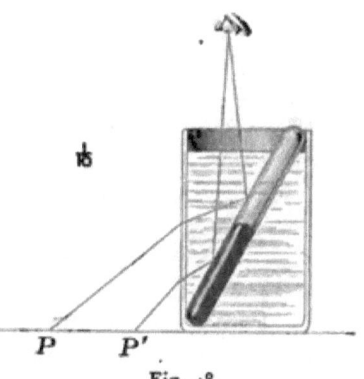

Fig. 48.

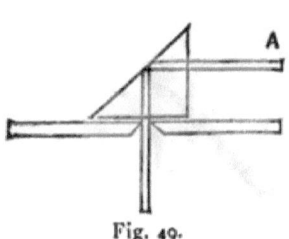

Fig. 49.

51. In explaining any of the facts or defining any of the terms referred to in this chapter you will find it best to do so with reference to the diagrams (at any rate if you wish to avoid the use of trigonometrical terms). But for the purpose of recapitulation the following definitions may here be given.

*Index of Refraction.*—If points be taken in the incident and refracted rays equidistant from the point of incidence, and if from these points perpendiculars be drawn to the normal, the ratio of the perpendiculars is constant for any pair of media, and is called the index of refraction.

*Critical Angle.*—When a ray of light is travelling from a denser towards a rarer medium, there is a certain angle of incidence such that the corresponding angle of refraction is 90°; and this angle of incidence is called the critical angle for the two media. *Or*—The critical angle is that limiting angle of incidence which just allows a ray travelling in the denser medium to escape into the rarer one—the refracted ray grazing the surface of separation of the two media.

*Total Reflection* is the reflection of light from the surface of separation of two media when the incident ray is in the denser medium, and the angle of incidence is greater than the critical angle.

Observe that *two* conditions are necessary before total reflection can occur:—

(1) The incident ray must be travelling from a denser medium towards a rarer one.

(2) The angle of incidence must be greater than the critical angle.

**52. Alternative Construction for Refracted Ray.**—Let IC (Fig. 50) be a ray travelling in water and incident upon the surface at C.

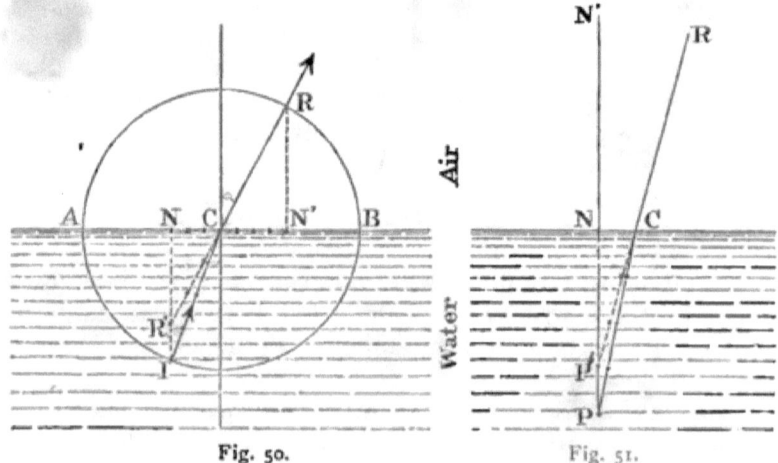

Fig. 50.    Fig. 51.

Let IN be drawn perpendicular to the surface, and let CR be the direction of the refracted ray in air, as found by the construction given in Art. 47.

Produce RC backwards to meet IN at R.' The triangles CNR' and CN'R are similar: for the opposite angles at C are equal, and the angles at N and N' are right angles. Hence the sides about the equal angles at C are proportional.

Now $$CN = \frac{3}{4} \cdot CN',$$

and, therefore, $$CR' = \frac{3}{4} \cdot CR = \frac{3}{4} \cdot CI.$$

This gives us another construction for finding the path of **the refracted ray.** Let IC be the incident ray. Draw IN perpendicular to the surface of separation. Divide CI into four equal parts. Along IN find **a point** R' such that CR' is equal **to three of** these parts. Join R'C and produce it: **this is** the path of the refracted ray **in air.** We shall use this construction in the next article.

**53. Image of Point under Water.**—Refer back to Fig. 47. Suppose any two consecutive **rays in the figure to** represent the extreme rays of a pencil of light proceeding **from the point A and** entering **an** observer's eye after refraction **into the air.** Draw a similar figure on a larger scale: **produce the refracted rays** backwards, and determine the point at which they intersect in the water. This point is the image of A. You will find that it differs for different pairs of rays: **the** position of **the virtual** image depends upon the position of the observer's eye.

When the observer is at a great distance **the image** is very near **the surface.** As he moves nearer the image moves farther down. **The motion of the image** finally becomes very slow, and when the **observer's eye is** vertically above **the** object, the image is formed on the vertical line at a depth below the surface **equal** to three-fourths of the depth of the object.

For let P (Fig. 51) be a luminous point under water. A ray of light PNN' incident **normally passes through into the air** without deviation, and the image of P **is formed on** this **vertical line.** The rays by which the observer sees **this image** are not all **vertical but form a small** divergent pencil. Let PC be any one of the **rays refracted along CR.** Produce RC backwards **to meet the** normal at P'. Then, according to the last article,

$$CP' = \frac{3}{4} \cdot CP.$$

But, since the pencil of rays **entering the eye is very** small, C very nearly coincides with N and R with N'. Hence we have, approximately,

$$NP' = \frac{3}{4} \cdot NP.$$

**All** nearly normal **rays** proceeding from P will, after refraction, appear to proceed from P'. **P'** is the virtual image of P as seen by an eye vertically above.

Suppose P to be **a** point at the bottom of a pond: to an **observer** looking vertically down, the depth of the pond would only appear to be three-fourths of what it really is. When the line of vision is oblique, the apparent depth is still smaller. Thus a pond or stream of uniform depth appears to be deepest vertically beneath the observer; and the bottom presents a wavy appearance, the trough of the wave being just beneath him.

For **the** same reason, a speck **on the underside** of a thick glass slab **appears nearer than it really is.** In general, the apparent thickness of a layer of any transparent medium is $\frac{t}{\mu}$, when $t$ is its real thickness, and $\mu$ the index of refraction for light travelling from air into the medium.

EXPT. 31.—Focus a microscope upon any well-defined object, such as a metal scale, small print, etc. Place a thick sheet of glass upon the object and look through again. You will find that it is now out of focus, and that the tube of the microscope has to be *raised* in order to focus the object again.

The effect of interposing a stratum of water can be observed by pouring water into a flat-bottomed glass standing on a printed page.

**54. Refraction through a Plate.**—By a plate is here meant a portion of a refracting medium contained between two parallel planes.

Let QA (Fig. 52) be a ray of light incident obliquely at A upon such a plate, *e.g.* a sheet of glass. On entering the plate at A the ray is refracted towards the normal along AS and falls upon the second face of the plate at S. Since the two faces of the plate are parallel, the angle of incidence of the plate at S is equal to the angle of refraction (BAS) at A. Hence (see Arts. 47, 48) the angle of refraction (into air) at S must be equal to the angle of incidence at A, and the emergent ray ST *is parallel to the incident ray* QA.

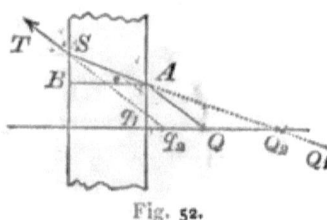

Fig. 52.

At the same time it must be noticed that the ray in passing through the plate is shifted to one side or laterally displaced. The amount of this lateral displacement increases with the thickness of the plate, its refractive index, and its inclination to the incident ray.

EXPT. 32.—Draw a few straight lines on a sheet of paper, and place a piece of plate-glass on the paper so as to cover a portion of the lines. Looked at normally through the glass the lines still appear straight (Fig. 53). But if you move your eye to the right or left the lines appear broken

Fig. 53.—NORMAL VIEW.   Fig. 54.—OBLIQUE VIEW.
Lateral displacement produced by glass plate.

at the edge of the plate (Fig. 54). The parts seen through the glass are still straight, but are shifted in the same direction as your eye.

The same effect is observed when the eye is kept fixed and the plate itself is tilted.

EXPT. 33.—The above experiment can be repeated with liquids, using a parallel-sided glass trough such as that shown on p. 90, and affords a simple method of proving that different liquids *have different refractive indices*.

Half fill the trough with water and hold it obliquely in front of a straight line drawn on a vertical sheet of paper : observe the lateral displacement.

Now fill the trough up with turpentine, which will float on the surface of the water. The line appears broken just where the water and turpentine meet, the part seen through the turpentine being *more* displaced than the part seen through the water. From this it follows that turpentine must have a higher refractive index than water, for the thickness and the inclination are the same for both liquids.

**55.** It can be shown by experiment that the direction of a ray emerging from a plate is parallel to that of the incident ray. We might have started with this as an experimental fact, and then have *deduced* from this the statement made in Arts. 47, 48, viz. that the index of refraction from any medium into air is the reciprocal of the index of refraction from air into that medium.

The interposition of a plate of glass not only causes a lateral displacement, but also makes an object appear *nearer* than it really is. Thus to an observer looking at a point S (Fig. 55) through a glass plate, the apparent position of the point is at S'—to one side and nearer. This latter result is not only produced when the incidence is oblique, but also when it is normal (see Art. 53).

Fig. 55.—REFRACTION THROUGH A PLATE.

EXPT. 34.—The effect of interposing a glass plate in any position may be shown as in Expt. 31.

**56. Multiple Images.**—Refraction plays a part in the production of the multiple images described in Art. 25. Suppose a ray of light from a source S (Fig. 56) to fall upon a plate-glass mirror. Part of the light is reflected from the front surface of the glass, and produces an image which is seen in the direction of S'. Part of it is refracted into the glass and then reflected from the silvered surface : of this, a part is again refracted into the air, producing the bright image seen in the direction of S°. But another part is reflected back into the glass, and after further reflection from the back of the plate and refraction into air proceeds parallel to the

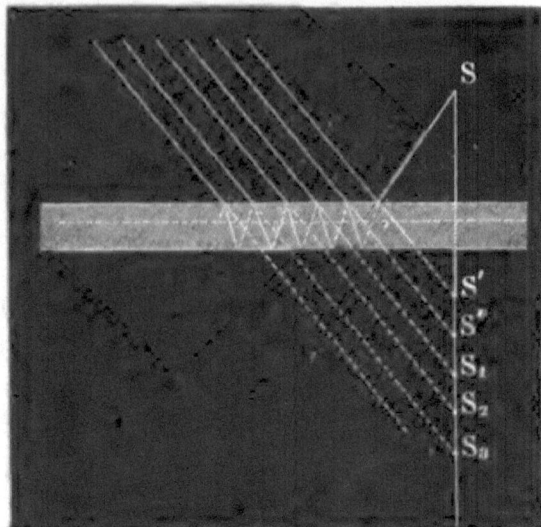

Fig. 56.—MULTIPLE IMAGES: PATHS OF RAYS.

other rays and produces the image $S_1$. Thus a series of images of gradually decreasing brightness is produced.

**57. Refraction through Prisms.**—A prism is a wedge-shaped portion of a refracting medium contained between two plane faces. The angle between the faces is called the *refracting angle* of the prism, and the line along which the faces meet is called the *edge* of the prism. Fig. 57 represents an equilateral glass prism (the section of which is an equilateral triangle).

Let *abc* (Fig. 58) be a section of a prism of which the refracting angle

Fig. 57.

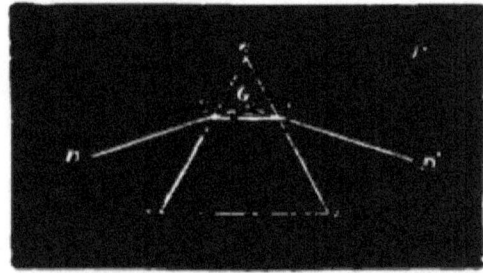

Fig. 58.

is at *a*. We shall suppose the prism to be of glass or other substance optically denser than air. A ray such as DE incident upon one face of the prism at E is refracted *towards* the normal and travels through the prism in the direction EE'. On leaving the prism at E' the ray is passing from a denser to a rarer medium: hence it is refracted *away from* the normal along E'D'. Thus, in our figure, the ray at each refraction is turned away from the edge of the prism or towards its base.

Produce DE and D'E' to meet. The angle FGD' is called the **angle of deviation**: it is the angle between the directions of the incident and emergent rays, and measures the total deviation produced by the prism. The deviation is always *away from the edge of the prism.*

The glass **pendants attached to** chandeliers are usually triangular prisms, and can be used for illustrating the behaviour of prisms. The effect of interposing a prism between your eye and an object is to shift the apparent position of the object *towards* the edge of the prism. Thus an observer at D (Fig. 58), looking through a prism at an object D', would see a virtual image of it at F.

**58. Minimum Deviation, etc.**—The deviation produced by a prism depends not only upon its angle and refractive index, but also upon its position with reference to the direction of the incident light. It can be proved that the deviation is least (or a *minimum*) when the angles of incidence and emergence are equal. When this is the case, the path of the ray within the prism is equally inclined to the two faces (as in Fig. 58), and the position of the prism in which this occurs is called the '*position of minimum deviation.*' This is the position in which prisms are usually placed in optical experiments.

In the case of prisms of small angle (say up to $15°$) the deviation is approximately proportional to the angle of the prism. This should be borne in mind when we come to consider how a lens acts (Art. 63).

EXPT. 35.—On a table or drawing-board place a prism with its edge vertical. Beyond the prism stick a **pin** (to represent a source of light). Look through the prism and note the apparent position of the pin; this depends upon the position of the prism. Twist the prism slightly, first in one direction and then in the other: **find** by trial the position in which the deviation is a minimum.

By using four pins—two on one side of the prism **and two on** the other—you can fix the directions of the incident and emergent rays and then *measure* the deviation. (When the pins are properly adjusted they should all appear to be in a straight line.) Do this with two prisms of the same angle but of different materials, and observe that the deviations are different: *e.g.* a prism of flint-glass produces a greater deviation than one of crown.

**59.** Objects **seen** through prisms generally exhibit coloured edges: the cause of this will be explained in Chap. VIII.

Several of the experiments **described in the present** chapter are suitable for projection with the lantern, *e.g.* Expts. 31-34.

The effect of heat in altering the **refractive** index of liquids and gases has been referred to in pp. **90 and 93.**

### Examples on Chapter VI

1. Explain the apparent bending of a stick when dipped into water, stating broadly from experience the most favourable conditions for observing the effect.

2. Light falls at a given angle on a plane refracting surface, for which the refractive index is 5/4. Show, by a geometrical construction, drawn, as well as you can, to scale, how to find the path of the refracted ray. (Apply the construction given in Art. 46.)

3. A ray of light passes from one medium into a second, the angle of incidence being 60°, and the angle of refraction 30°: show that the index of refraction is $\sqrt{3}$.

4. The critical angle for a certain medium is 45°: show that its index of refraction is $\sqrt{2}$. (See Art. 49.)

5. Draw accurately the path of a ray of simple light through a 45° prism of glass, whose index of refraction is $\frac{5}{3}$, drawing the ray incident on one face in a direction perpendicular to the other face.

6. The shadow of a red-hot poker is cast on a white screen by means of a lime-light lantern. Explain the smoky appearance on the screen just above the shadow.

7. Explain the quivery appearance seen above hot bricks or rocks, and the streaky appearance of water in which ice or sugar is being dissolved.

# CHAPTER VII

## LENSES

**60.** A **lens** is a portion of a refracting medium bounded either by two curved surfaces or by one plane surface and one curved surface. The only lenses which we shall consider are glass lenses of which the curved surfaces are portions of spheres. We shall further assume that the greatest thickness of the lens is small compared with the radii of curvature of the surfaces.

**61. Kinds of Lenses.**—Lenses may be divided into two classes :—

I. Those which are thicker at the centre than at the edge. These are called **convex or converging lenses**. Sections

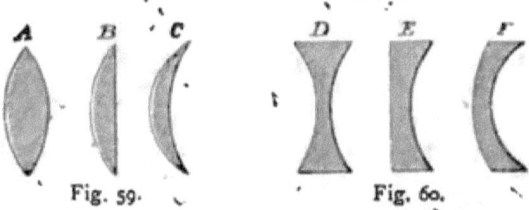

Fig. 59.    Fig. 60.

of three lenses of this kind are shown in Fig. 59. Of these A has two convex surfaces and is called a *double-convex* lens ; B has one plane and one convex surface and is called a *plano-convex* lens ; while C is called a *converging meniscus*.

II. Those which are thinner at the centre than at the edge. These are called **concave or diverging lenses**. Three lenses of this kind are shown in Fig. 60. Of these D is a *double concave* and E a *plano-concave* lens, while F is a *diverging meniscus*.

**62. The principal axis** of a lens is the line joining the centres of curvature of its two spherical surfaces. The axis of a plano-convex or plano-concave lens is the line which passes through the centre of curvature of the spherical surface and is perpendicular to the plane surface.

A ray of light travelling along the axis of a lens falls normally on both refracting surfaces, and therefore passes through the lens without suffering any deviation.

**63. How a Lens Acts.**—In any *convex* lens the inclination of the two faces towards one another increases as we go outwards from the centre (or axis) of the lens towards the edge. Thus we may imagine the section of the lens to be made up of a number of prisms of gradually increasing angle, as shown in the accompanying diagram.

We know that a ray of light in passing through a prism is deviated towards its base, and that the amount of the deviation increases as the angle of the prism increases. Now suppose a beam of parallel rays to fall upon the prism-lens, shown in Fig. 61. The rays would be bent towards the axis, those near the edge being deviated more than those nearer the centre. The result would be to convert the parallel beam into a convergent pencil.

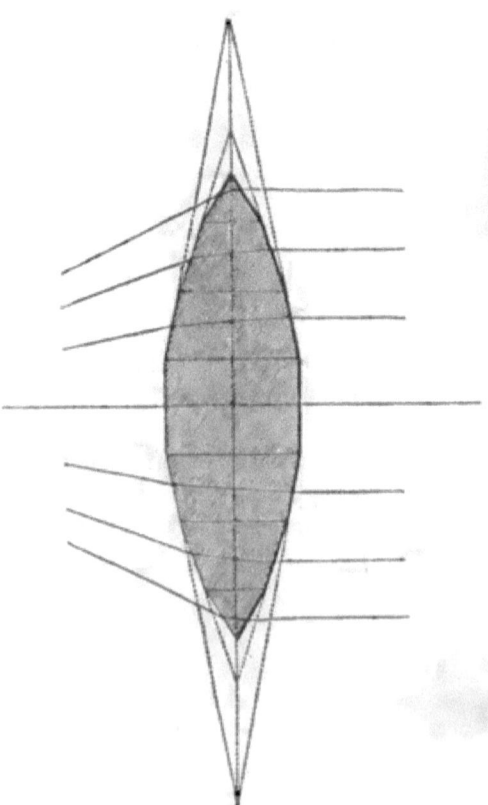

Fig. 61.

This is the way in which convex lenses act. Their general effect is *to render transmitted rays more convergent.*

The section of a *concave* lens may be regarded as being built up of a number of prisms of gradually-increasing angle, arranged with their bases outwards (or away from the centre). The general effect of such lenses is *to render transmitted rays more divergent.*

Thus the properties of convex lenses are similar to those of concave mirrors ; while the properties of concave lenses are similar to those of convex mirrors.

**64. Optical Centre of a Lens.**—The axis is not the only direction in which a ray of light can pass through a lens without suffering deviation. There is one point on the axis of every lens such that, if the path of a ray within the lens passes through this point, the emergent ray is parallel to the incident ray. This point is called the **optical centre** of the lens, and any line passing through it is called a *secondary axis* of the lens. In the two accompanying diagrams of convex and concave lenses O is the optical centre, and the line IOI′ is a secondary axis. At the points I and I′ where this line meets the lens its two surfaces are parallel to one another, and consequently the lens acts upon a ray traversing it in this direction just as if it were a parallel-sided plate: the ray emerges parallel to its direction before incidence, but suffers a slight lateral displacement (Art. 54).

The optical centre is sometimes simply referred to as the centre of the lens. It is not necessarily the geometrical centre: indeed in the case of a meniscus it may be altogether outside the lens. But in the common types of lenses shown in Fig. 62 (double-convex and double-concave lenses with surfaces of equal curvature) the optical centre *does* coincide with the geometrical centre, or is midway between the points A and B.

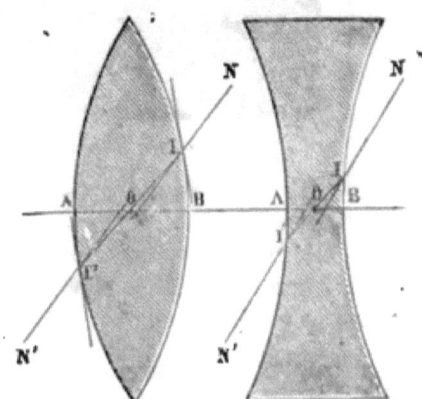

Fig. 62.—OPTICAL CENTRE.

O

Without explaining how the position of the optical centre is determined, we may define it as follows with reference to its most important property:—

*The optical centre of a lens is a point on the principal axis, and is such that, if the path of a ray within the lens passes through this point, the emergent ray is parallel to the incident ray.*

Assuming the thickness of the lens to be small, we may neglect the slight lateral displacement produced and regard *any ray through the optical centre as passing straight through the lens without deviation.* We may further neglect the thickness of the lens altogether and regard O as coinciding with either A or B (Fig. 62).

**65. Principal Focus.**—When a pencil of rays parallel to the principal axis of a convex lens falls upon the lens, the rays after transmission through it converge to a point on its axis (F, Fig. 63). This point is called the **principal focus.**

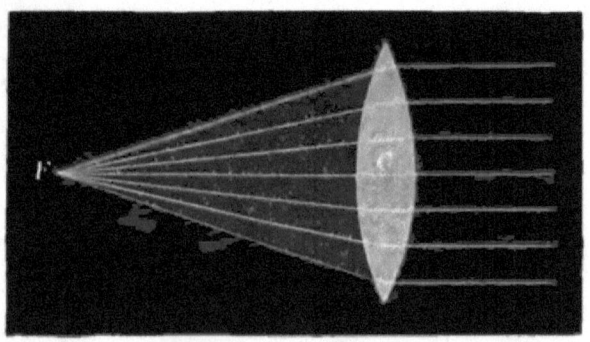

Fig. 63.—CONVEX LENS: PRINCIPAL FOCUS.

Since the rays actually pass through the point, the focus is clearly *real*.

In the case of a concave lens the transmitted rays diverge so as to *appear* to come from the principal focus (F, Fig. 64), which is therefore *virtual*.

**66. Focal Length.**—The principal focal distance or **focal length** of a lens is the distance between the principal focus and the lens itself (or, more strictly, its optical centre). This is usually denoted by the letter $f$.

Applying the rules for signs given in Art. 30, it will be seen that the focal length of a *convex* lens is *negative*, while that of a *concave* lens is *positive*. For all distances have now to be measured from the lens (or its optical centre). Now in the case of the convex lens (Fig. 63) the source of light is on the right hand of the lens, and the principal focus is on the left hand. Hence the focal length (CF) is negative (−).

In the case of the concave lens (Fig. 64) the principal focus is on the same side as the source of light (the right hand), and therefore the focal length is positive (+).

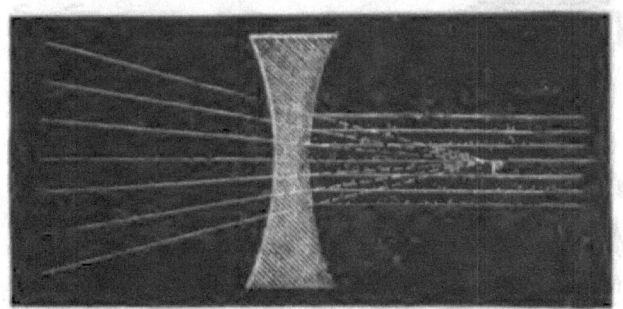

Fig. 64.—CONCAVE LENS: PRINCIPAL FOCUS.

**67. Parallel Beam.**—We may always suppose the direction in which a ray of light travels to be reversed: it will then exactly retrace its original path. Applying this principle of reversal to the convex lens (Fig. 63), we see that, if the path of an incident ray passes through the principal focus, the ray will, after transmission, go off parallel to the principal axis.

Thus a parallel beam of light can be obtained by placing a luminous point at the principal focus of a convex lens. As a source of light we may use a small brightly-illuminated hole in a cardboard screen (Art. 37).

A more powerful beam of parallel light can be obtained from the optical lantern, the source of light being placed at the principal focus of the 'condenser,' which usually consists of a pair of plano-convex lenses. Convex lenses are used in the same manner in lighthouses, in signal lamps, and in the common 'bull's-eye' lanterns.

**68. Conjugate Foci.**—The following proposition can be proved by the laws of refraction to be very approximately true, and we shall presently see that experiment confirms it. When rays from a luminous point on the principal axis of a lens fall upon it, the transmitted rays either converge to another point on the axis or appear to diverge from such a point. In the former case (when the rays actually pass through it) the point is called a real focus: in the latter case (when they only appear to diverge from it) it is a virtual focus.

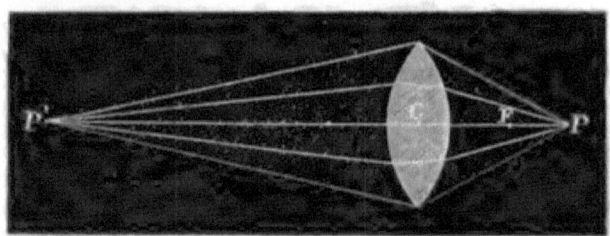

Fig. 65. CONJUGATE FOCUS (REAL).

In the case of a convex lens the focus is *real* when the source of light is *beyond* the principal focus of the lens, and is *virtual* when it is *nearer* to the lens than the principal focus. For we have seen that rays diverging from the principal focus of a lens are rendered parallel by transmission through it. Now if the source of light be at P (Fig. 65), beyond the principal focus (F), the divergence of the incident rays is less,

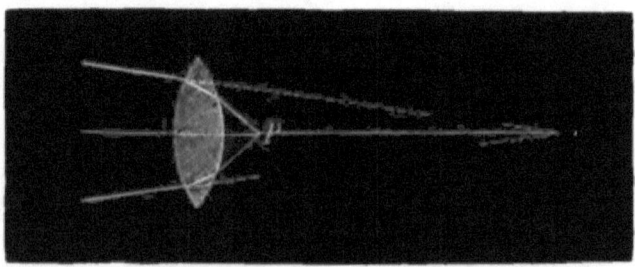

Fig. 66. CONJUGATE FOCUS (VIRTUAL).

and by transmission through the lens they are made to converge to a point P' on the other side.

If the source of light be nearer, as at *p* (Fig. 66), the divergence of the incident rays is greater than when the source is at the principal focus: thus the lens cannot even render them parallel, but it diminishes the divergence so that the transmitted rays appear to diverge from a point $p'$ on the same side of the lens but farther off.

In Fig. 65 P' is the image of P formed by the lens. Now, if you suppose the direction in which the light travels to be reversed, it will be clear that P is also the image of P'. The points P and P' are called **conjugate foci**, with respect to the lens: they are so related that *one is the image of the other as formed by the lens.* This may be taken as a definition of conjugate foci *when both are real.*

When one of the foci is *virtual* the case is somewhat different. Thus in Fig. 66 $p'$ is the image of $p$, whereas $p$ would *not* be the image of $p'$. But the principle of reversal still holds good: a pencil of rays *converging* to $p'$ would after transmission through the lens converge to $p$.

**69. Equation for Conjugate Foci.**—It can be proved that the distances of the conjugate foci P and P' are connected by the relation

$$\frac{1}{f} = \frac{1}{p'} - \frac{1}{p} \qquad \qquad (1)$$

where $f$ denotes the focal length of the lens, $p$ the distance of the point P (the object or source of light) from the lens, and $p'$ the distance of P' (the corresponding focus or image) from it.

This is the fundamental equation for lenses, and with the proper conventions as to signs (see Arts. 30 and 66) it holds good for both convex and concave lenses.

Compare this with the equation for mirrors given on p. 152. You will be less liable to confuse the two if you at once notice (1) that the present equation contains a minus ($-$) instead of a plus sign, and (2) that the term containing $p'$ comes first.

**70.** The farther a source of light is from the lens, the more nearly parallel are the rays which proceed from it to the lens. In fact we may regard parallel rays as being such as proceed from a source at an infinite distance. In this case $p$ is infinitely great and $\frac{1}{p}$ is infinitely small, or $=0$. Equation (1) now reduces to

$$\frac{1}{f} = \frac{1}{p'}, \quad \text{or } p' = f,$$

which is in accordance with what was stated in Art. 65. The image of an infinitely distant source of light (*e.g.* the sun) is formed at the principal focus of the lens.

**71. Examples.**—1. The focal length of a convex lens is 15 cm. Rays of light diverge from a point on the principal axis of the lens and 20 cm. in front of it: to what point will they converge after refraction through the lens?

Let $p'$ be the required distance of the conjugate focus from the lens. Since the lens is convex, its focal length is negative: thus $f = -15$, and $p = +20$. Inserting these values, *with their proper signs*, in the equation

$$\frac{1}{f} = \frac{1}{p'} - \frac{1}{p},$$

we get

$$-\frac{1}{15} = \frac{1}{p'} - \frac{1}{20}.$$

Thus

$$\frac{1}{p'} = \frac{1}{20} - \frac{1}{15}$$

$$= \frac{3-4}{60} = -\frac{1}{60},$$

or $p' = -60$ cm. The refracted rays converge to a point 60 cm. *behind* the lens (as in Fig. 65).

2. If the luminous point in the last example were 12 cm. from the lens, how would the rays behave after transmission through the lens?

Here $p = +12$ and $f$ as before $= -15$. Inserting these values in the equation, we get

$$-\frac{1}{15} = \frac{1}{p'} - \frac{1}{12}.$$

Thus

$$\frac{1}{p'} = \frac{1}{12} - \frac{1}{15}$$

$$= \frac{5-4}{60} = \frac{1}{60},$$

and $p' = +60$ cm. The conjugate focus is on the same side of the lens as the luminous point. This is what might have been expected, for the source of light is at a smaller distance from the lens than its focal length. This case corresponds to that illustrated in Fig. 66: the transmitted beam is divergent and appears to proceed from a *virtual* focus 60 cm. from the lens and on the same side as the incident light.

3. Rays of light diverging from a point 6 in. before a lens are brought to a focus 18 in. behind it: what is the focal length of the lens?

Here $p = +6$ and $p' = -18$. The value of $f$ is given by the equation

$$\frac{1}{f} = -\frac{1}{18} - \frac{1}{6}$$

$$= \frac{-1-3}{18} = -\frac{4}{18} = -\frac{2}{9},$$

and
$$\therefore f = -\frac{9}{2} = -4\frac{1}{2}.$$

The focal length is $-4\frac{1}{2}$ in. Since it is negative, the lens must be convex.

## Convex Lenses

**72. Geometrical Construction for Image.**—Images produced by convex lenses may be either real or virtual. In discussing the formation of images by lenses we shall make use of geometrical constructions similar to those employed in the case of mirrors (see Arts. 32, 33, and 35).

Among the many rays which proceed from any luminous point to the lens, there are three whose directions after refraction are easily followed:—

(1) Any ray whose path passes through the optical centre may be regarded as passing straight through the lens (Art. 64).

(2) Any ray parallel to the principal axis is refracted so as to pass through the principal focus.

(3) Any ray whose path passes through the principal focus will, after refraction through the lens, proceed parallel to its principal axis.

If we wish to find the image of any luminous point, we need only follow out the directions of two of the above-mentioned incident rays—say (1) and (2)—and find the point of intersection of the refracted rays: this is the image of the luminous point. If the refracted rays actually intersect at this point, the image is real: if they have to be produced backwards (in front of the lens) before they intersect, the image is virtual.

**73. Real Images.**—Let AB (Fig. 67) be an object in front of a convex lens whose optical centre is at O and principal focus at F. OF is the direction of the principal axis.

Join AO and produce it: the image of A lies somewhere

along this line (which is the secondary axis through A). From A draw a second incident ray parallel to the principal axis:

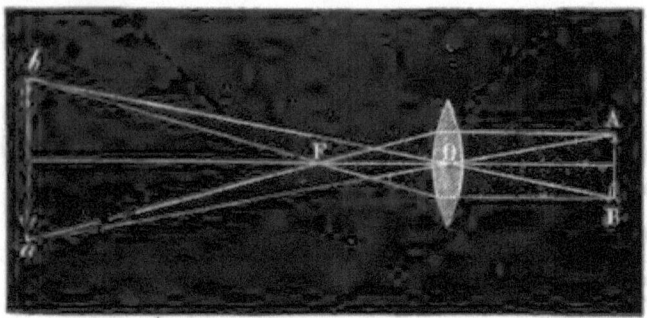

Fig. 67.—REAL, INVERTED, AND ENLARGED IMAGE.

the corresponding refracted ray passes through F. The image of A lies somewhere along this ray: it must therefore be formed at *a*, where the two refracted rays intersect.

Similarly the image of B is found to be at *b*. The images of points lying between A and B are formed at corresponding points between *a* and *b*. *ab* is the image of AB.

The image is *real*. Not only can it be seen by an eye placed in a suitable position, but it can also be thrown on a screen (see Expt. 37).

In Fig. 67 the image is enlarged, but this is not always the case. We shall see in the next article that the size of the image depends upon the distance of the object from the lens, and that it may be either enlarged or diminished. This will also be evident from the following consideration. Whenever a real image is formed, the image and object may change places. Thus if *ab* in Fig. 67 were an object, AB would be its image, and in this case the image would clearly be smaller in size than the object.

The object and its real image are always on opposite sides of the lens; and, since the rays which pass through the centre (O) of the lens cross each other at this point, the real image is always *inverted*.

**74. Relative Sizes of Image and Object.**—In order to find the relative sizes of image and object, we take as our object the straight line AB (Fig. 68) drawn perpendicular to the principal axis BCF of the lens. The image of A is at $a$, the position of which is found by the usual geometrical con-

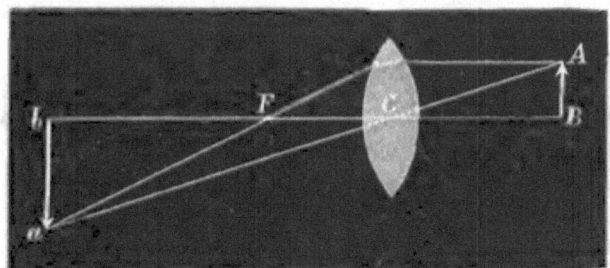

Fig. 68.

struction. The image is completed by drawing $ab$ perpendicular to the principal axis (the image of B being at $b$ on the axis).

The triangles $ab$C and ABC are similar; for the opposite angles at C are equal and the angles at $b$ and B are right angles. Hence (Euclid, VI. 4) the sides about these angles are proportionals, *i.e.*,

$$\frac{ab}{AB} = \frac{bC}{BC}.$$

Now $b$C is the distance of the image from the optical centre of the lens. If we suppose the thickness of the lens to be small, we may say that $b$C is the distance of the image from the lens, and BC is the distance of the object from it. Thus

*The relative sizes of image and object are as their respective distances from the lens.*

If we denote the sizes of image and object by I and O, and their respective distances from the lens by $p'$ and $p$, we may state the result in the form

$$\frac{I}{O} = \frac{p'}{p} \qquad . \qquad . \qquad . \qquad . \qquad (2)$$

75. The above result is true for all images formed by lenses. When we come to consider virtual images and concave lenses, you should follow out the proof for yourself in each case. Notice also that here the relative sizes are expressed directly in terms of distance from the lens. In the

case of mirrors (Art. 34) the proof was not so simple. The diagram (Fig. 33) gave the relative sizes in terms of the distances from the centre of curvature, and then we had to prove that the sizes were also proportional to the distances from the mirror itself.

The ratio $\left(\dfrac{I}{O}\right)$ between the sizes of image and object is called the *magnification*. Here, as usual, the word 'size' refers to *linear* dimensions.

When an object is placed at a distance $2f$ in front of a convex lens of focal length $f$, a real image is formed at an *equal* distance behind the lens, and the image and object *are of the same size*.

For example, suppose an object to be placed 60 cm. in front of a convex lens of 30 cm. focal length. To find the position of the image we have to put $p = +60$ and $f = -30$ in the equation

$$\frac{1}{f} = \frac{1}{p'} - \frac{1}{p}.$$

Thus
$$-\frac{1}{30} = \frac{1}{p'} - \frac{1}{60},$$

or
$$\frac{1}{p'} = \frac{1}{60} - \frac{1}{30} = -\frac{1}{60},$$

and
$$\therefore p' = -60.$$

The image is formed 60 cm. *behind* the lens. Since the image and object are equally distant from the lens, they are also equal in size.

When the distance of the object from the lens is *greater* than $2f$, the image is *diminished*. When the distance of the object is *less* than $2f$, the image is *enlarged*. But no real image is formed when the distance of the object is less than $f$ (see Art. 79).

**76. Examples.**—1. An object is placed at a distance of 3 ft. in front of a lens and the image is formed 1 ft. behind the lens. What is the focal length of the lens? and what kind of lens is it?

Here $p = 3$ and $p' = -1$.

Thus
$$\frac{1}{f} = \frac{1}{p'} - \frac{1}{p},$$
$$= -\frac{1}{1} - \frac{1}{3} = -\frac{4}{3},$$

and
$$f = -\tfrac{3}{4}.$$

The focal length is $\tfrac{3}{4}$ ft. or 9 in., and since it is negative the lens is convex.

2. *An object whose length is* 5 *cm. is placed at a distance of* 12 *cm. from a convex lens of* 8 *cm. focal length: where is the image formed and how long is it?*

Here $f = -8$ (since the lens is convex) and $p = +12$.

The distance of the image from the lens is given by the equation

$$-\tfrac{1}{8} = \tfrac{1}{p'} - \tfrac{1}{12},$$

or

$$\tfrac{1}{p'} = \tfrac{1}{12} - \tfrac{1}{8},$$

$$= \tfrac{2-3}{24} = -\tfrac{1}{24}.$$

Thus $p' = -24$. The image is formed at a distance of 24 cm. on the other side of the lens. Since its distance from the lens is twice that of the object, its length is also twice that of the object, or $= 10$ cm.

3. *I have a convex lens of* 15 *cm. focal length. How far from an object must I hold it so that it may produce a real image of the object magnified three times?*

Suppose $p$ and $p'$ to be the required distances of the object and image from the lens. Since the image is to be three times the size of the object, its distance from the lens must be three times as great, or $p' = 3p$ (numerically).

It must, however, be noticed that in stating the relation

$$\tfrac{I}{O} = \tfrac{p'}{p} \qquad\qquad . \quad . \quad . \quad . \quad (2)$$

in Art. 74, we took no notice of the *signs* of the quantities $p$ and $p'$. Since our image is to be real it must be formed on the *opposite* side of the lens (as in Fig. 68); thus $p'$ is *negative* and $= -3p$.

In the usual equation

$$\tfrac{1}{f} = \tfrac{1}{p'} - \tfrac{1}{p},$$

we have now to put $f = -15$ and $p' = -3p$. We thus get the equation

$$-\tfrac{1}{15} = -\tfrac{1}{3p} - \tfrac{1}{p},$$

in which $p$ is the only unknown quantity. Solving this we find that $p = +20$. Thus the lens must be held at a distance of 20 cm. from the object. The image is formed at a distance of 60 cm. ($3 \times 20$) on the other side of the lens.

[In working out questions of this kind, a diagram drawn roughly to scale is a useful guide. You can also check your work by finding out whether the values of $p$ and $p'$ found are in accordance with the value of $f$ given.]

**77. Experimental Illustrations.**—The student should now perform the following experiments with a convex lens, and then proceed to measure carefully the focal length of several convex lenses by the methods described in Art. 78. The large (front) lenses of an opera-glass or telescope will do: but the best plan will be to get from an optician a few cheap unmounted lenses of focal lengths varying from 2 in. to 1 ft. (say 5-30 cm.) These can be mounted in wood or cork mounts and clamped to a retort-stand. For purposes of measurement, the lens, candle, and screen are best mounted on the optical bench (p. 129).

EXPT. 36.—Focus the sun's rays on a sheet of paper by means of a strong magnifying-glass. The bright round spot of light is an image of the sun, and is formed at the principal focus of the lens (Arts. 66-69). The focal length of the lens is the distance between it and the image.

Strong convex lenses are often called 'burning-glasses,' because they concentrate heat as well as light at the focus. Bits of paper and chips of wood are easily set on fire by focussing the sun's rays on them. We have thus a proof that 'radiant heat,' like light, can be refracted.

EXPT. 37.—Place a lighted candle at one end of a dark room. In the middle of the room, and at the same height as the flame, place a convex lens. Move a screen backwards and forwards on the other side of the lens until the real inverted image of the flame is sharply defined on it. Proceed as in the case of the concave mirror (p. 160, Expts. 19 and 20). Notice particularly the following points.

The image is real, inverted, and diminished. If the distance of the candle is great compared with the focal length of the lens, the image is formed very nearly at the focus. If you lower the candle, the image rises, and *vice versa*: the image of a point always lies on the secondary axis passing through that point.

As the candle moves up towards the lens, the image moves away from the lens and becomes gradually larger. Measure the distances of the object ($p$) and image ($p'$) from the lens in several positions, and see whether $\frac{1}{p'} - \frac{1}{p}$ remains constant and equal to $\frac{1}{f}$. (See Art. 69. Remember that $p'$ is negative here. If you focus the image carefully, your results should agree with the theory.)

When the candle and screen are at equal distances from the lens, the image is equal in size to the flame itself. The distance of each from the lens is $2f$, or the focal length is one-quarter the distance between the candle and its image when both are equal in size (Art. 75).

As the candle moves still nearer the lens the image moves still farther away: it is now larger than the object, but remains real and inverted. This holds good for distances of the object between $2f$ and $f$.

When the flame is at the principal focus it gives a parallel beam or forms an image at an infinite distance. When it is still nearer the lens ($p<f$) no real image is produced, but on looking at the flame through the lens you see a virtual, erect, and enlarged image.

**78. Methods of Finding Focal Length.**—The focal length of a convex lens can be measured by any of the following methods:—

I. By allowing a beam of parallel rays—or rays from any very distant object—to fall on the lens. An image of the object is formed at the principal focus (see Expt. 36).

II. The above method can be modified by making use of the lens itself to form the parallel beam.

Suppose F (Fig. 69) to be a source of light placed at the principal focus of a convex lens C. After passing through the lens the rays of light would form a parallel beam. Now let a plane mirror MM' be held normally in this beam, so as to reflect the light back. Each ray would travel back along its own path and would again be refracted by the lens, so as to pass through its focus F. Thus an image coincident with the source of light would be produced.

This gives us a simple and accurate means of measuring the focal length of the lens. It is similar to Method III for concave mirrors, and is carried out with the apparatus there described (p. 162). SS' is a cardboard screen provided with hole and cross-threads. The hole is brightly illuminated by

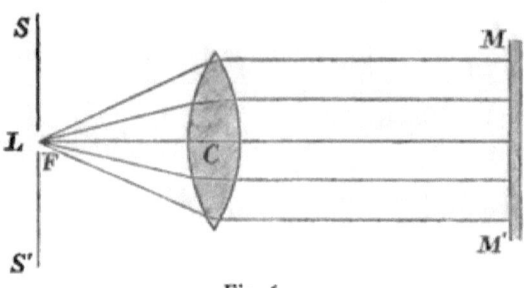

Fig. 69.

an Argand burner placed behind (at L). The lens is placed at a distance from the screen roughly equal to what the focal length is supposed to be. The exact distance of the mirror from the screen is not important: it should be tilted so that the blurred spot of light seen on the screen should fall just by the side of the hole.

Now move the lens backwards or forwards until this blurred spot becomes a bright image of the hole, with the cross-threads well defined. The distance between the screen and the lens is now the focal length.

III. By throwing a real image of an object on a screen, and measuring their respective distances from the lens. The focal length can then be calculated from the equation

$$\frac{1}{f} = \frac{1}{p'} - \frac{1}{p}.$$

As an object we may use a candle-flame or the hole and cross-threads above described. Measurements should be made with the object at different distances. The value of $f$ should be calculated from each experiment, and the mean value taken as correct.

IV. Find by trial that position of the object which gives an equal-sized real image at an equal distance on the other side of the lens. Measure the distance between the object and the image: one-fourth of this is the focal length. This is simply a special case of Method III.

## 79. Virtual Images.

Virtual images are produced by convex lenses when the distance of the object from the lens is less than its focal length.

In Fig. 70 the position of the focus is marked at F and $f$ on either side of the lens. The object AB is between the lens and its principal focus. O is the optical centre of the lens.

Join AO. The image of A lies somewhere along this line,

Fig. 70.—CONVEX LENS: VIRTUAL IMAGE.

which is the secondary axis through A. From A draw a ray parallel to the principal axis of the lens: this is refracted through the focus F. These two rays do not meet, however, if they are produced on the left-hand side of the lens. But if produced backwards they *do* meet at a point on *the same side* of the lens as the object, and this point is the virtual image of A. The image of B is found in the same way, and lies on the secondary axis OB.

An observer looking at the object AB through the lens sees a virtual, erect, and enlarged image of it. Thus a convex lens can be used as a 'magnifying-glass' or simple microscope for examining small objects or reading small print.

The magnification depends upon the relative distances of the image and object (Art. 74). These distances are given by the usual equation for conjugate foci (see examples in Art. 81).

## 80. Table of Results.

In the following table $f$ denotes the focal length of the lens, which is supposed to be convex. Distances are measured from the lens.

| Distance of Object. | Distance of Image. | Characteristics of Image. |
|---|---|---|
| Infinite | $f$ | Real, inverted, diminished. |
| Greater than $2f$ | Between $f$ and $2f$ | Real, inverted, diminished. |
| $2f$ | $2f$ | Real, inverted, equal in size. |
| Between $2f$ and $f$ | Greater than $2f$ | Real, inverted, enlarged. |
| Less than $f$ | Greater than $f$ (on same side as object) | Virtual, erect, enlarged. |

**81. Examples.**—1. **An object is** placed at a distance of **10 cm.** from a convex lens of **15** cm. **focal length:** find the position and characteristics of the image.

Here $f = -15$. If $p'$ is the distance of the image from the lens,

$$-\frac{1}{15} = \frac{1}{p'} - \frac{1}{10},$$

$$\therefore \frac{1}{p'} = \frac{1}{10} - \frac{1}{15} = \frac{3-2}{30} = \frac{1}{30},$$

and $p' = +30$. The image is formed at a distance of 30 cm. from the lens, **and** on the *same* side as the object. It is therefore *virtual* and *erect*. Since $p' = 30$ and $p = 10$, the image is *magnified* three times.

2. If an object at a distance of 3 in. from a convex lens has its image magnified three times, what is the focal length of the lens?

There are two solutions to this problem, for the image may be either real or virtual. In the first case the image and object are on opposite sides of the lens; in the second case both are on the **same** side of the lens.

In both cases, since the **image is three** times as large as the object, its distance from the lens must be three times that of the object: but this only gives us the *numerical* value of $p'$ in terms of $p$ without stating whether it is positive or negative.

(1) Image real—
$p'$ is negative and $= -3p$ or $= -9$ in.

Thus
$$\frac{1}{f} = \frac{1}{p'} - \frac{1}{p},$$
$$= -\frac{1}{9} - \frac{1}{3} = -\frac{4}{9},$$
$$\therefore f = -\frac{9}{4}$$

or the focal length of the lens is $-2\frac{1}{4}$ inches.

(2) Image virtual—
   $p'$ is positive and $= 3p = +9$ in.

Thus
$$\frac{1}{f} = \frac{1}{9} - \frac{1}{3} = -\frac{2}{9},$$

and
$$f = -\frac{9}{2},$$

so that the focal length in this case is $-4\frac{1}{2}$ inches.

### CONCAVE LENSES.

The general properties of concave lenses have already been discussed in Arts. 65 and 66.

**82. Images produced by Concave Lenses are always Virtual, Erect, and Diminished.**—Let AB (Fig. 71) be the object, O the optical centre of the lens, and F its focus.

A ray proceeding from A parallel to the principal axis will, after refrac-

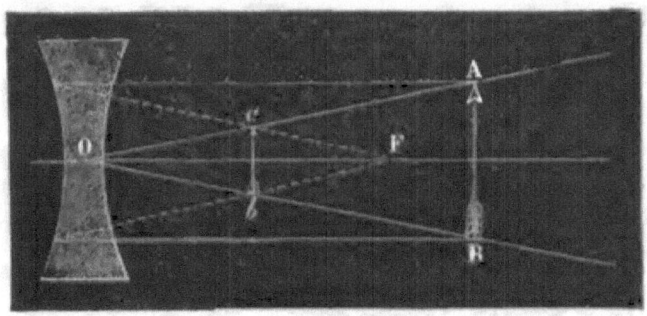

Fig. 71.—CONCAVE LENS: VIRTUAL IMAGE.

tion through the lens, appear to come from F in the direction of the dotted line. This line intersects the secondary axis AO at $a$. Rays proceeding from A emerge from the lens as though they came from the point $a$. $a$ is the virtual image of A. Similarly $b$ is the virtual image of B.

The image $ab$ is virtual, erect, and diminished. This is not only the

P

case when the object is beyond F (as in our figure), but is always true for all images formed by concave lenses.

**83. Examples.**—1. The focal length of a concave lens is 3 inches. Rays of light diverge from a point on its axis, and 6 in. from it. How will they proceed after refraction through the lens?

Here $p = +6$ and, since the lens is concave, $f = +3$. The distance of the conjugate focus is given by the equation,

$$\frac{1}{3} = \frac{1}{p'} - \frac{1}{6}.$$

Thus
$$\frac{1}{p'} = \frac{1}{3} + \frac{1}{6} = \frac{3}{6},$$

and
$$\therefore p' = +2.$$

The conjugate focus is at a distance of 2 in. from the lens. Since it is on the same side (+) as the source it is virtual. The rays after refraction proceed as though they diverged from a point 2 inches from the lens.

2. A virtual image of an object 30 cm. from a lens is formed on the same side of the lens and at a distance of 10 cm. from it. What kind of a lens is it?

The focal length of the lens is given by the equation,

$$\frac{1}{f} = \frac{1}{10} - \frac{1}{30} = \frac{2}{30},$$

and
$$\therefore f = +15.$$

Since the focal length is positive, the lens must be *concave*. This may also be seen from the fact that the image is nearer the lens than the object is, and is therefore smaller: if the lens were convex, the virtual image would have been magnified.

### EXAMPLES ON CHAPTER VII

Read carefully the **rules and** instructions **given in Arts.** 30, 65, 66, and 68-70. Observe how **these** are applied in the solved examples in Arts. 71, 76, 81, and 83. Draw diagrams **roughly to** scale **for** the purpose of guiding and checking your algebraical **work.** When you are about to substitute the numerical value for any symbol in an equation, be careful to give it the right sign (+ or −).

Remember that—

The focal length **of a convex** lens is *negative*.

**The** focal length of a concave lens is *positive*.

When $p'$ is found to be positive, the image **is formed on** the same side of the lens as the object, and is *virtual*; when $p'$ is **negative** the image is formed on the opposite side of the lens and is *real*.

The size of the image is to that of the object as $p'$ is to $p$. But if the image is real, $p'$ is **negative**: if it is virtual, $p'$ is *positive*.

1. **Rays of** light diverging from **a** point 1 ft. **in** front of a lens brought **to a** focus 4 inches behind it. What is the nature of the lens, and what is its focal length?

2. The focal length of a convex lens is 9 cm. Rays of light diverge from a point on its axis and 12 cm. in front of it. Find the position of the conjugate focus.

3. Rays of light diverge from a point 20 cm. in front of a convex lens. The focal length of the lens is 4 cm. How do the rays behave after refraction through it?

4. If an object is placed 1 ft. from a convex lens of 9 inches focal length, where is the image formed?

5. An arrow 4 inches long is placed 10 inches in front of a convex lens whose focal length is 4 inches. Illustrate by a figure, and find the position and length of the image.

6. An object is placed at a distance of 60 cm. from a convex lens of 15 cm. focal length: where is the image formed? Compare its size with that of the object.

7. An object whose length is 5 cm. is placed at a distance of 12 cm. from a convex lens of 8 cm. focal length: what is the length of the image?

8. A candle is placed at a distance of 10 ft. from a wall, and it is found that when a convex lens is held midway between the candle and the wall, a distinct image is produced upon the latter. Find the focal length of the lens, and the relative sizes of the object and image.

9. A coin half an inch in diameter is held 1 ft. in front of a convergent lens: if the focal length of the lens is 8 in., find the position and magnitude of the image.

10. Draw figures, approximately to scale, showing the paths of the rays of light, and the positions of the images formed when a luminous object is placed at a distance of (1) 1 inch, (2) 6 inches from a convergent lens of 2 in. focal length.

11. An object is placed 8 in. from a convex lens, and its image is formed 24 in. from the lens on the other side. If the object were placed 4 in. from the lens, where would the image be?

12. A candle stands at a distance of 2 metres from a wall, and it is found that when a lens is held half a metre from the candle a distinct image is produced upon the wall: find the focal length of the lens, and also state the relative sizes of image and object.

13. A lens of 9 in. focal length is to be used for the purpose of producing an inverted image of an object magnified three times: where must each be situated?

14. A convex lens is held 5 ft. from a wall: and it is found that there is one position in which an object can be held so that a real image of it, magnified five times, is thrown upon the wall. Determine this position, and also the focal length of the lens.

15. You are provided with a convex lens of 18 in. focal length, and are required to place an object in such a position that its image will be magnified three times: find the positions which will give (1) a real, and (2) a virtual image of the required size.

16. A person looks at an object through a concave lens of 1 foot focal length, the object being 5 feet beyond the lens. Draw a figure showing the paths of the rays by which he sees the image formed, and determine its position.

# CHAPTER VIII

## OPTICAL INSTRUMENTS

**84. The Optical Lantern.**—In its best-known form—the 'magic-lantern'—this instrument is mostly used for projecting upon a screen images of pictures photographed or painted upon glass. These are called 'slides.' The slide, held in a suitable slide-holder (*s* Fig. 72) is placed just in front of the condensing lenses *cc*, which are so called because they condense upon it the light emitted by the source *l*. The slide

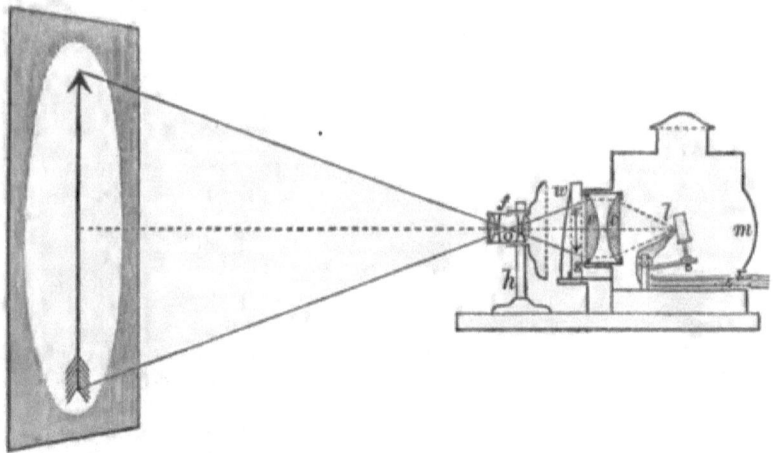

Fig. 72.—OPTICAL LANTERN AS USED FOR PROJECTION.

thus forms a brightly-illuminated object, in front of which is placed the 'objective' or projecting lens. This is generally a compound lens, but we may regard it as a simple lens of which the optical centre is at *o*. The objective produces a real, inverted, and magnified image of the slide. This image is focussed on the screen by sliding the lens-holder *h* backwards or forwards : a finer adjustment can be made by the focussing-screw *f*.

If the image on the screen is to appear right-side up, the slide must clearly be put in upside down.

**85. A Photographic Camera** or *camera obscura* consists of a small dark chamber, the body of which is generally made of folded leather or cloth, after the manner of a bellows, so that it can be shortened or lengthened at will. The front consists of a wooden board in which is mounted a lens. This throws an image of external objects upon a sheet of ground-glass let into the back of the camera. When the photographer is about to take a portrait or view he focusses this image sharply upon the ground-glass screen by moving either the back of the camera or the lens front. The screen is then replaced by a sensitised plate, which is usually a plate of glass coated with a film of gelatine containing finely-divided salts of silver.

**86. The Eye.**—From an optical point of view the human eye may be regarded as a camera obscura in which the glass screen or sensitive plate is replaced by a sensitive membrane (*rrr*, Fig. 73) called the *retina*. It contains a double-convex lens — the *crystalline lens*, *o* — which divides the eye into two chambers. The posterior chamber is filled with a jelly-like substance called the vitreous humour (*vv*).

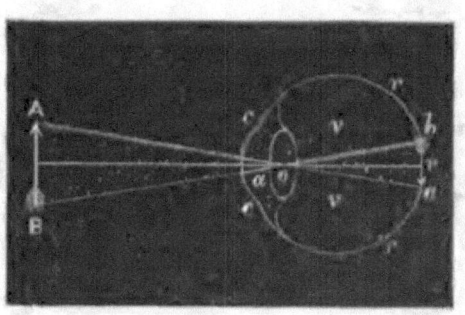

Fig. 73.

The anterior chamber is filled with a watery liquid called the aqueous humour (*a*) : this is bounded in front by the cornea (*cc*), a tough transparent membrane shaped like a very convex watch-glass.

Partly by refractions through these humours, but mainly by the action of the crystalline lens itself, images of external objects are formed upon the retina. For further explanation of the way in which the nerves are affected and the sensation of vision produced, the reader is referred to books on Physiology.

**87. Instruments for Magnifying.**—In Art. 79 we saw how a magnified virtual image of an object could be produced by means of a convex lens placed at a distance from the object less than its focal length.

When greater magnifying powers are required the following method is used. Let AB (Fig. 74) be the object and O a convex lens which produces a real image of it at *ab*. This image can not only be thrown on a screen or viewed directly

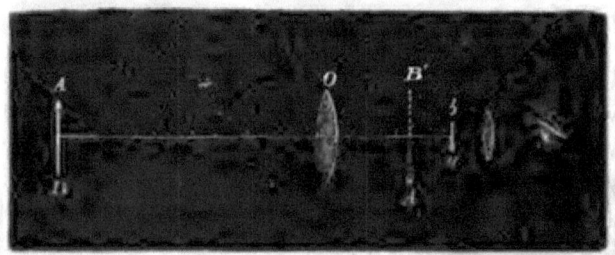

Fig. 74.

by an eye placed behind it, but it can also be observed by means of a second convex lens. For this purpose the lens must be placed at a distance from *ab* less than its focal length. An observer looking through the second lens sees at A'B' a magnified virtual image of *ab*. Instead of examining the object itself with a magnifying-glass he applies the magnifier to an image of the object.

This principle is employed in the construction of compound microscopes and astronomical (refracting) telescopes. The image seen is evidently inverted, but in the instruments named this causes no great inconvenience.

EXPT. 38.—Select two convex lenses for carrying out the method above described. Use as your object a pair of cross-wires, or a piece of ground-glass on which a small arrow has been drawn with a fine hard pencil. Illuminate the object brightly by a gas-flame.

The lenses are most easily adjusted as follows. Focus on a second ground-glass screen the real image produced by the first lens. Mark its position with a pencil on the glass. Focus the second lens on this mark by moving it backwards or forwards until a magnified virtual image of the mark is seen on looking through the lens. Now remove the glass and you will see a magnified and inverted image of the object.

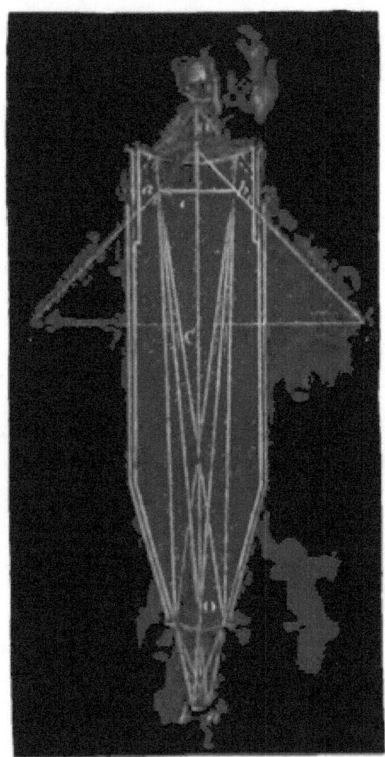

Fig. 75.

**88. Astronomical Telescope.**—This consists essentially of two lenses: (1) a large lens of considerable focal length called the *object-glass*, and (2) a smaller lens (or system of lenses) of short focal length called the *eye-piece*. When the telescope is directed towards a star, a real, diminished, and inverted image of the star is formed at the focus of the object-glass. This image is observed through the eye-piece, just as if it were an actual object, and thus a second and magnified image is obtained.

**89. The Compound Microscope** differs

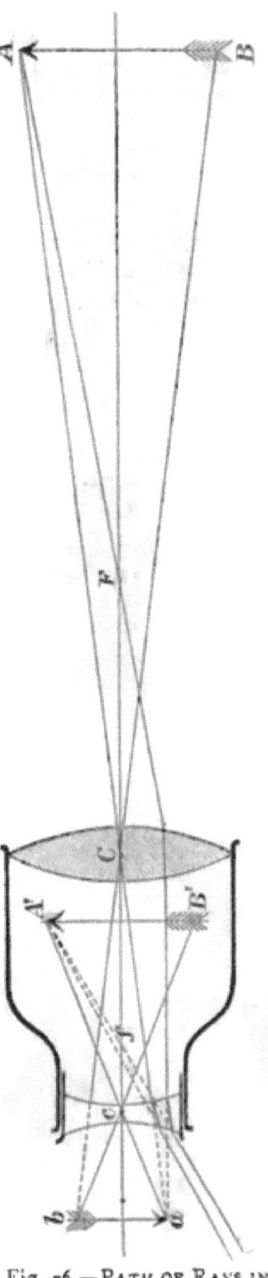

Fig. 76.—Path of Rays in Galileo's Telescope or Opera-Glass.

from the telescope in having a small object-glass of short focal length. By placing this at a suitable distance from the object a magnified real image is obtained. **The** path of the rays within the microscope is shown in Fig. 75. The object-glass O produces at $a_1b_1$ a real image of the object *ab*. This image, already magnified, **is** viewed through the eye-piece O' and **thus** the observer sees at **AB a** virtual and greatly magnified image.

**90. Galileo's Telescope—Opera-Glass.**—In the earliest form **of** telescope, **invented** by Galileo, the object-glass consists as usual **of a** convex **lens, but** the eye-piece is a concave lens. The way in which **the** lenses **act** will be understood from Fig. 76. The object-glass C alone would **form** at *ab* a real, **inverted** image of the object AB. But when the concave lens *c* is interposed, each convergent pencil is converted into a divergent pencil. Thus, **the** rays which would otherwise converge to the point *a* are made, **by passing** through the concave lens, to diverge from a **virtual focus at A'** (A' being on the secondary axis *ac*). Thus **the observer sees at A'B' a virtual** image of the distant object AB.

An ordinary opera-glass **is** simply a double-barrelled Galilean telescope. This form of telescope has **the** advantage of producing an erect image.

**EXPT.** 39.—Start as **in Expt.** 38 ; but instead of using a second convex lens **take a concave** one. Interpose it between the first (convex) lens and the **image on the** screen. Look through the concave lens and adjust its position **until you see distinctly a** virtual, erect image of the arrow.

# CHAPTER IX

## DISPERSION AND COLOUR

**91.** In experimenting with prisms, and in examining the images produced by lenses, you have doubtless already noticed the appearance of certain coloured edges. These are seen even when the object itself is not coloured: indeed they are most distinct when the object is white or when it consists of a hole or slit illuminated by white light. White light in fact is not simple but **compound**; and by refraction it is split up into its component parts. The composite nature of white light was discovered by **Newton**.

**92. Newton's Experiment.**—Newton allowed a beam of sunlight to stream through a small round hole in the window-shutter of a darkened room. In the path of the beam he placed a glass prism so as to refract the light upwards towards the opposite wall. But instead of seeing on the wall a round or oval image of the sun he found that by passing through the prism the light was drawn out into a coloured band of considerable length, violet at the top and red at the bottom (Fig. 77). This he called the spectrum. The colours of the spectrum follow the

Fig. 77.

same order as the tints of the rainbow; they pass imperceptibly from red at the one end through all the gradations of orange, yellow, green, and blue, to violet at the other end.

**93.** We learn from this experiment:—

(1) That white light is not simple but compound: it is the result of a mixture of many colours.

(2) That these colours can be separated by passing the light through a prism.

(3) That various colours have various degrees of refrangibility; violet being the most refrangible and red the least.

**94. Dispersion.**—Our second law of refraction (as stated on p. 172) is only correct so long as we keep to light of one colour, say yellow light. On entering a refracting medium a violet ray is bent more than a yellow ray, and therefore the index of refraction for violet light is somewhat greater than the index for yellow light; for red light it is less than for either violet or yellow. The difference between the amounts of bending produced by passing through a prism produces a separation or *dispersion* of the various coloured rays.

**95.** Newton also placed a second prism behind the first with its edge in the same direction, so as to cause a further refraction. He found that no new colour was introduced; the second prism simply increased the amount of deviation or lengthened the spectrum.

He then turned the second prism round, placing its edge towards the base of the first prism, so that the refractions due to the two prisms were in opposite directions. He now found that there was thrown upon the wall a white image of the sun (slightly displaced, as it would be by passing through a thick plate of glass). We learn from this that the various spectral colours mixed in the proper proportions can be made to *recombine and form white light.* · Other methods of performing this recomposition are given in Expts. 43 and 44 (p. 222).

**96. A Narrow Slit necessary.**—In Newton's experiments the light was admitted through a round hole. Now, if you imagine such a hole to be divided up into narrow strips, in a direction parallel to the edge of the prism, you will easily see that each strip of light would produce a spectrum of its own:

these successive spectra would overlap each other, so that the colour in each part of the resultant spectrum would be mixed up with some of the colours lying on each side of it. For this reason the colours in Newton's spectrum were not *pure*. The first requisite for the production of a *pure spectrum* is that the light should come through a narrow slit parallel to the edge of the prism. In Art. 98 it will be shown that certain other adjustments are necessary for projecting a pure spectrum on a screen. For the present we shall assume that the spectrum is observed directly by looking through a prism at a narrow slit or narrow strip of white paper pasted on a black background.

**97. Colour due to Absorption.**—Now let a piece of coloured glass, say red glass, be placed in front of the slit. The more refrangible part of the spectrum (violet, blue, etc.) is cut off: only the red and orange get through. By interposing the red glass we do not introduce into the spectrum any colour which did not exist in it before: hence we conclude that the glass appears red not because it introduces any fresh colour into the light which is transmitted through it, but because it *stops or absorbs certain other colours*.

A similar explanation holds good for the colours exhibited by bodies when they are viewed, as is more usual, by the light which they reflect. For, in general, the reflection does not take place wholly from the surface of the body: some of the light penetrates a little distance below the surface, and, in so doing, its colour may be modified by the suppression of other colours. A rose appears red because it reflects mainly red light and absorbs the more refrangible end of the spectrum; and its leaves appear green because they contain a green colouring-matter (chlorophyll) which possesses the property of absorbing red light. A lily appears white because it reflects equally all the component colours of the sunlight. Viewed in red light it would simply appear red; and in blue light, blue.

**98. Production of a pure Spectrum.**—We have already seen in Art. 96 that a narrow slit is required for the production of a pure spectrum. In order that the spectrum may be sharp and distinct it is also necessary to place the prism in the position of minimum deviation (Art. 58), for this is the position in which the best definition is secured. This is all that is necessary when the spectrum is observed by looking directly through the prism. But if the spectrum is to be thrown upon a screen something further is required.

For let S (Fig. 78) be the slit. Through this there streams a wedge-shaped beam of light which, before the prism is interposed, throws upon the screen a patch of white light $ll'$. When the prism is interposed, the beam of light suffers refraction and dispersion. If it consisted altogether of red light, there would be seen upon the screen a red patch of light at $rr'$; and if it consisted of violet light there would be a violet patch at $vv'$.

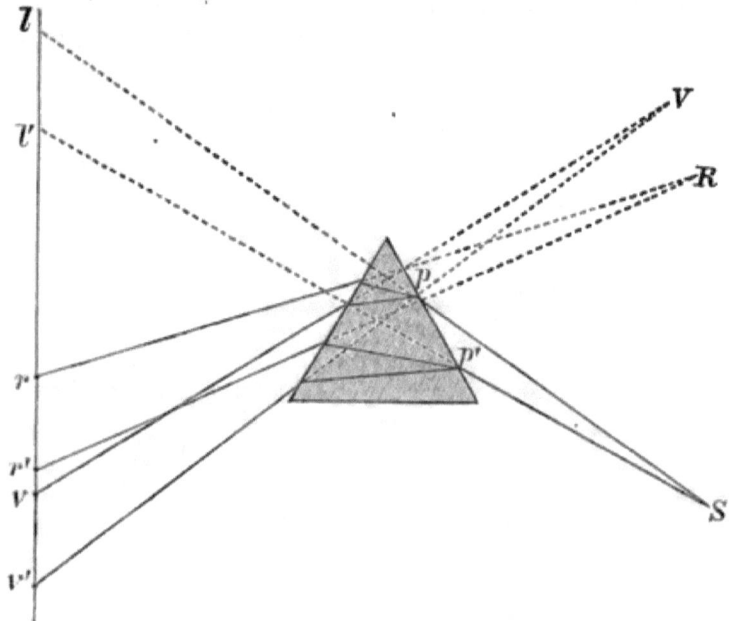

Fig. 78.—IMPURE SPECTRUM FORMED BY PRISM.

All the red light, after passing through the prism, would appear to *diverge* from a virtual focus at R, and all the violet light from a virtual focus at V. An observer looking through the prism in this direction would see at RV a virtual spectrum of the white light proceeding from S.

But upon the screen itself the spectrum is confused and impure, because the various coloured patches overlap each other. This overlapping would be prevented if we could convert each patch into a narrow strip, and this we can do by interposing a convex lens as shown at L (Fig. 79). The lens converts the divergent beam into a *convergent* beam, and, if the prism were absent, would produce at S' a real image of the slit. If the light were all red, the lens and prism would together produce on the screen a real image of the slit at R; and if it were violet they would produce a real image at V. Thus, with white light, we obtain upon the screen a series of sharply-defined coloured images of the slit arranged side by side and forming a pure spectrum.

We may therefore state as follows the *three conditions requisite for the production of a pure spectrum*:—
(1) The slit must be narrow.
(2) A real image of the slit must be formed by a convex lens.
(3) The prism must be placed in the position of minimum deviation.

**99. Experiments with the Spectrum.**—For projecting a spectrum on a screen it is best to make use of an optical lantern (Art. 84) furnished with the lime-light. All stray light should be excluded from the room.

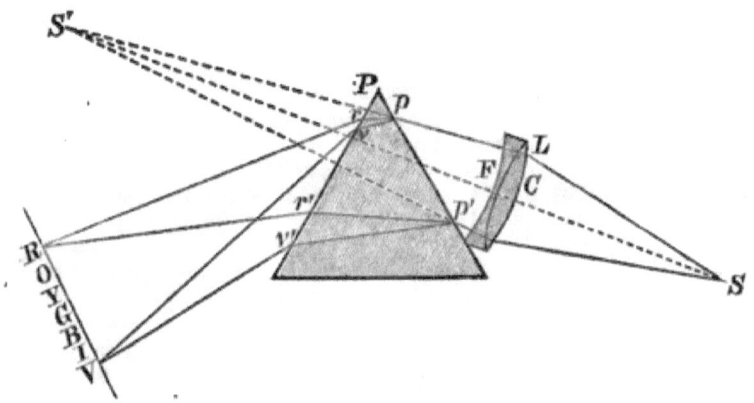

Fig. 79.—Pure Spectrum formed by Lens and Prism.

The nozzle of the lantern is to be covered with a blackened cap, as in Expt. 12, but in this case the slit (about 3 millimetres wide) is to be vertical. A fairly large lens, of about 25 cm. focal length, should be used. The most suitable prisms are hollow glass prisms filled with carbon bisulphide [$CS_2$]; this substance has a higher dispersive power than glass, and therefore gives a longer spectrum than a solid glass prism does. The spectrum should be received on a screen somewhat larger than itself (say 70 cm. long and 50 cm. high) and made of thick white paper stretched on a wooden frame.

EXPT. 40.—In the path of the lantern beam, and about 2 metres from the slit, fix a small screen. On this, by means of the lens, focus a real and enlarged image of the slit (as at S', Fig. 79): if the lens has the focal length above stated it will have to be put about 30 cm. from the slit. Behind the lens place the prism. Place the large screen in a suitable position for receiving the spectrum, as at RV, Fig. 79, and let its distance from the prism be equal to the measured distance of S' from the prism. Adjust the prism so that the deviation is a minimum. This can be done by trial—twisting the prism slightly in one direction or the other until the red end R of the spectrum on the screen is as near as possible to S'. You now have upon the screen a pure and bright spectrum of the light radiated by the hot lime. With a 60° bisulphide prism, and adjustments made as above, the spectrum will be about 40 cm. long.

EXPT. 41.—Hold various coloured glasses in front of the slit and observe what parts of the spectrum are cut out. Notice that scarcely any of the transmitted colours are *pure*. Red glass lets through a fairly pure red mixed with some orange: but blue (cobalt) glass not only lets through the blue and violet end of the spectrum but also a considerable amount of red.

Examine in the same way coloured solutions contained in flat glass cells. A strong solution of potassium bichromate cuts off all the more refrangible end of the spectrum, letting through only red, orange, and yellow. An ammoniacal solution of copper sulphate lets through the green, blue, and violet only. Thus the absorptions of these solutions are complementary, and if they are both placed together in front of the slit they stop all the light.

EXPT. 42.—Pass a blue ribbon through the spectrum. In the more refrangible part, especially the blue, it appears of its ordinary colour. In the red end it appears black; it absorbs all the light that falls upon it and reflects none to the eye. A bright red ribbon shows the effect even more distinctly: placed in the orange of the spectrum, one edge of the ribbon appears red and the other edge (in the yellow) appears black. Thus the colour of a body depends not only upon the nature of the body itself but also upon the nature of the light which falls upon it.

EXPT. 43.—If you have another prism of the same kind you can repeat Newton's experiment on the recomposition of white light by placing the second prism close behind the first and edge to base. You will thus get again a white image of the slit on the small screen at $S'$ (Fig. 79).

Instead of a prism you may use a lens for recombining the colours. The best kind of lens for this purpose is a cylindrical lens; but it is not necessary to get one of glass, for a large beaker filled with water answers perfectly.

EXPT. 44.—The effect of mixing colours can also be studied by means of circular discs of cardboard divided up into coloured sectors. In order to show the recomposition of the various spectral colours, the tints and sizes of the sectors should be chosen so as to correspond as nearly as possible to the tints and breadths of the various colours in the spectrum. If such a disc is attached to a whirling-table or a humming-top, and made to rotate rapidly in a good light, it appears of a uniform whitish-grey.

### EXAMPLES ON CHAPTER IX

1. Describe an experiment proving that white light is compound. How can it be shown that the constituents into which it is resolved are not likewise compound?

2. One piece of glass appears dark red, and another appears dark green, when held up to the light ; explain why, when they are both put together, no light can be seen through them.

3. A ribbon purchased by gaslight appeared to match the dress with which it was to be worn. Next morning the match appeared to be very imperfect. Explain this.

# ANSWERS TO EXAMPLES

### Chapter II (p. 129)

**5.** As 4 is to 9.  **6.** As 9 is to 4.  **7.** 10·24 candle-power.  **8.** 1 ft. from the candle and 3 ft. from the lamp.

### Chapter V (p. 167).

**3.** The image would be formed 1 ft. 4 in. in front of the mirror.  **4.** 6 cm. *behind* the mirror.  **5.** $f = +30$ cm.: the mirror is concave.  **6.** The image is 7 ft. 6 in. in front of the mirror and is 5 in. long.  **7.** $f = +12$ cm.: the mirror is concave.  **8.** The image is 3 ft. from the mirror and 3 in. long.  **9.** The image is 10 in. in front of the mirror and $2\frac{2}{3}$ in. high.  **10.** $f = +8$ in.  **11.** $p = +18$ in.; $p = +36$ in. The image would be one-half the size of the object.  **12.** Real, $33\frac{1}{3}$ cm. in front of the mirror, and 4 cm. long.  **14.** 30 in. behind the mirror; magnification $= 6$.  **15.** The image moves through 3·7 in.; *i.e.* from 24 in. behind the mirror to 20·3 in. behind it.  **17.** (1) Image is virtual and erect, 4 in. behind the mirror, 2 in. long; (2) image is 8 in. behind the mirror, 1 in. long.  **18.** The image is 7·5 cm. behind the mirror and 3 cm. long.

### Chapter VII (p. 210)

**1.** Convex: $f = -3$ in.  **2.** 36 cm. behind the lens.  **3.** They converge to a point 5 cm. from the lens and on the other side.  **4.** 3 ft. from the lens on the other side.  **5.** Image is real, 6 in. long and $6\frac{2}{3}$ in. from lens.  **6.** $p' = -20$ cm. : one-third size of object.  **7.** 10 cm.  **8.** 2 ft. 6 in. : equal in size.  **9.** $p' = -2$ ft. : diameter of image $= 1$ in.  **11.** 1 ft. from the lens and on the same side as the object.  **12.** $f = -37 \cdot 5$ cm. ; as 3 to 1.  **13.** The object must be 1 ft. from the lens and the image 3 ft. (on the opposite side).  **14.** $p' = +12$ in. ; $f = -10$. in.  **15.** (1) $p = 2$ ft. ; (2) $p = 1$ ft.  **16.** $p' = +10$ in.

# SOUND

# CHAPTER I

## INTRODUCTION—VIBRATORY MOTION

**1. Sound caused by Motion.**—If you examine any sounding body you will soon come to the conclusion that the sound is produced by motion of some kind. In many cases the motion is obvious and can be followed by the eye. For example, if you take hold of a stretched string between your finger and thumb, pluck it to one side and then let it go, the string will give out a sound and, at the same time, will be seen to swing from side to side. If the rate of motion is very rapid you may not be able to follow each separate swing, but the fact that the string is in motion is shown by its losing its definite outline and presenting the appearance of an indistinct gauzy spindle. By touching the string with the finger the motion can be stopped, and it is then found that the sound ceases at the same time.

The same indistinct appearance is presented by the prongs of a tuning-fork while it is sounding. When any hard object, such as the point of a knife, is held against the prongs, a loud rattling noise is heard; and quite a little shower of spray is thrown up when the points of the fork are dipped into water. If you want further proof of the motion you can easily get it by striking the fork and then making it touch your lips or teeth.

**2. The kind of Motion.**—A tuning-fork consists essentially of two steel springs united together at one end, and each prong of the fork moves in much the same way as a single straight spring clamped at one end. By taking a long thin spring (instead of the comparatively short and thick prong of the tuning-fork) we can make the motion slower and examine it at our ease.

EXPT. 1.—Clamp the lower end of the spring in a vice. Pull the upper end to one side, as at *l*, Fig. 1. The spring is *elastic*: it bends or 'gives.' In order to bend it you have to exert *force* upon it. In virtue of its elasticity the spring tends to recover its original form, and if you wish to keep it bent you must continue the pressure against it.

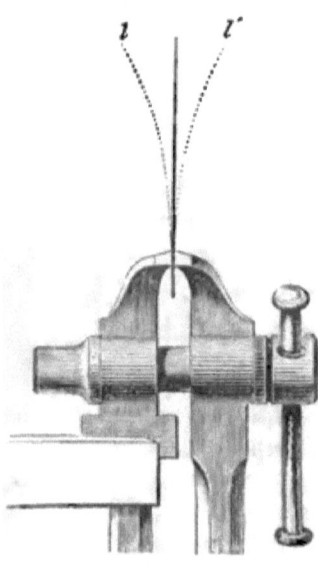

Fig. 1.

Now let the spring go. It flies back: but it does not stop when it has got back to its original position (vertical). It overshoots the mark and moves, with gradually diminishing velocity, to about an equal distance on the opposite side (*l'*). It then begins to return, again overshoots the mark, gradually comes to rest, starts back again, and so goes on swinging from side to side. Owing to the resistance of the air, etc., the amplitude of the excursions gradually diminishes, and finally the spring comes to rest.

**3. Vibration.**—When a body or point moves in the manner above described it is said to *vibrate* or *oscillate*. Referring again to Fig. 1, a movement from *l* to *l'* and back again to *l* is called a *complete vibration*. The *amplitude* of the vibration is measured by the extent of the motion on either side of the mean position or position of rest, *i.e.* half the distance *ll'*.

If you take a watch and examine the rate of vibration of the spring, you will find that this rate is regular and *independent of the amplitude of vibration*. Whether the excursions made by the end of the spring are great or small, the number of vibrations executed in a given time is constant. Thus the vibration is regular or *periodic*. The time required to perform a complete vibration is called the *period* of vibration. We shall denote this time (measured in seconds or a fraction of a second) by $\tau$. If the number of complete vibrations performed in a second be denoted by $n$, then

$$\tau = \frac{1}{n}.$$

We may call $n$ the *frequency* of vibration or the *vibration-number*.

**4. Vibration of Pendulum.**—The periodic vibration of an elastic spring is very similar to that of a pendulum swinging under the action of the earth's attraction. When the bob of the pendulum is displaced to one side and then let go, it swings to a nearly equal distance on the other side, stops, returns, and goes on swinging from side to side regularly, but with the amplitude of the oscillations gradually diminishing, and finally the pendulum comes to rest. The vibrations are regular: the period of vibration is constant and independent of the amplitude. As in the case of the spring, each complete vibration may be divided up into four similar parts.

The pendulum is in its mean position, or position of rest, when the string is vertical and the bob is at its lowest point (N, Fig. 2). Now suppose the bob to be displaced to one side, say to A. The bob not only moves to the left but rises, for it moves in an arc of a circle. In order to raise it through the distance HA you have to exert force and do *work* in lifting the weight of the bob upwards. You have thus given to the bob the power of doing an equal amount of work in falling through the same distance, *i.e.* to its position of rest. It possesses energy of position, or *potential energy*.[1] When the pendulum is released it moves with gradually increasing velocity, and this velocity is greatest when the bob is at its lowest point. The energy has now been changed from the potential into the *kinetic* form. It is this kinetic

---

[1] The *energy* of a body is its capacity for doing **work**.

A body is said to possess *potential energy* when **it is able to** do work in virtue of its *position*. A raised weight (as in a clock) and a head of water are examples of bodies possessing potential energy. So also are coiled springs, compressed air, etc. ; so that 'position' must here be held to include change of form or volume.

A **body is** said to possess *kinetic energy* when it is able to do work in virtue of its *motion*. **Thus** the wind does (useful) work in turning the sails of a wind-mill; a **cannon-ball does** (destructive) work in crashing through the walls of a fort.

energy possessed by the bob that prevents it from coming to rest at its lowest point and carries it over to the other side (A'), at the same time lifting it up against the action of gravity. At the end of the first swing (or semi-oscillation) the bob comes to rest for an instant and then retraces its path. In the first quarter-period the energy changes from the potential to the kinetic form, and in the second quarter-period from the kinetic to the potential form. In the third and fourth quarters the same changes are gone through with velocities reversed.

Observe that the velocity is least (zero) at the points where the displacement is greatest (A and A') : the velocity is greatest when the displacement is zero (N).

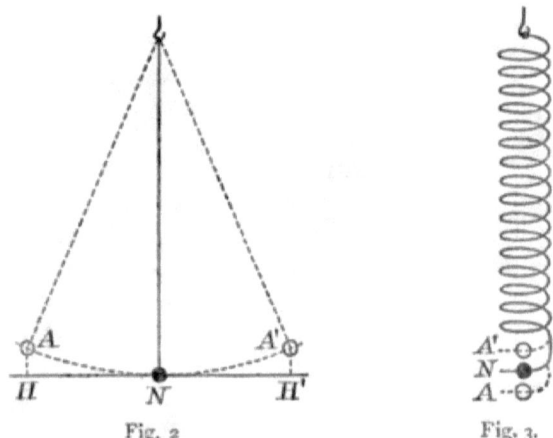

Fig. 2  Fig. 3.

5. **Simple Straight-line Vibration.**—If the distance through which a pendulum swings is small compared with the length of the string, the path of the bob is very nearly a straight line. The following is a better illustration of such vibration.

EXPT. 2.—Coil some thin wire round a pencil or cork-borer. Hang the spiral up and to its lower end attach a lead bullet (N, Fig. 3). Pull the bullet down to A and then let it go. It flies up to A' and continues for some time to oscillate up and down along a vertical line. This vibration along a straight line may be regarded as a type of the kind of motion which accompanies sound-waves in air.

**6. Elasticity.**—The force tending to bring the pendulum back to its position of rest is due to the attraction of the earth on the bob. In the case of the spring the force is due to the elasticity of the spring itself. When the spring is bent or pulled out, work is done in overcoming the resistance due to its elasticity, and this work is stored up as potential energy in the spring itself. After being released it moves in the same way and goes through the same changes of energy as the pendulum. Like the latter it is gradually brought to rest by resistances due to the air, etc.

The elasticity of a body may be defined as being *that property in virtue of which it requires force to change its form or volume, and recovers its original form or volume when the force is removed.*

Solids in general offer resistance to any change either of form or volume, and are therefore said to possess elasticity of form as well as elasticity of volume. Fluids exhibit the latter only, for they do not offer resistance to change of form (see p. 27).

In common language we speak of substances like india-rubber as being very elastic, because they 'give' or stretch readily. In scientific language the term is used in quite a different sense. Without entering into any details of the methods of measuring elasticity, we may here state generally that the elasticity of a substance is measured by the force required to produce a given change of form or volume. If the substance yields readily when force is applied, its elasticity is said to be small: but if it offers great resistance, it is said to be highly elastic. Thus the elasticity of steel and glass is great, for both these substances require the application of considerable force to produce even a small change of form. Again, the elasticity of water is very much greater than that of air: if the atmospheric pressure were to be doubled, the volume of any given quantity of air would (according to Boyle's law) be reduced to one-half; whereas such a change of pressure would produce scarcely any appreciable effect upon the volume of a given quantity of water.

**7. Graphical Representation of Vibration.**—The following experiment shows how the vibration of a tuning-fork may be studied and represented graphically.

EXPT. 3.—Smoke one side of a strip of glass by holding it over the flame of burning turpentine or camphor. Make a light pointed style of thin sheet-brass or wire: fix this to the end of one prong of a tuning-fork, as shown in Fig. 4. A bristle attached with wax may be used instead, but is apt to

Fig. 4.—Wave-Line traced by Fork.

bend or break off. Strike the fork, and immediately draw the style lightly but quickly over the smoked glass. After a little practice you will be able to produce a beautifully regular wavy line on the glass. This 'wave-line' is a record of the motion of the prong in its own handwriting.

**8. Musical Sounds and Noises.**—By musical sounds are meant such as are produced by the human voice, the violin, the organ, and other instruments used for musical purposes. It is not possible to draw any very sharp distinction between a musical sound and a noise. The sounds produced by certain instruments used in military bands—*e.g.* drums and triangles—cannot strictly be called musical, and yet they do not destroy the harmony of the music, but rather strengthen and brighten it.

We may, however, broadly distinguish between the two classes of sounds as follows. A noise is produced by confused and non-periodic movements,—movements which are irregular both in respect of time and strength: whereas a musical sound is produced by regular and periodic vibrations. Our business lies entirely with the latter class of sounds, and we shall in general use the word sound as denoting a musical sound.

**9. Pitch.**—The long thin spring used in Expt. 1 gave out no sound while vibrating. A certain rate of vibration must be reached before any audible sound is produced. By gradually shortening the spring you can make it vibrate more rapidly, until at last it begins to produce a deep sound or a note of 'low pitch.' By still further shortening the spring (or by using a shorter and thicker one) you can go on increasing the rate of vibration: this will make the note still higher or 'raise the pitch.' Thus the pitch of a note depends upon the rate of vibration. Further proofs of this will be given later on.

**10. Intensity of Sound.**—The intensity or loudness of a sound depends upon the amplitude of vibration of the sounding body. You can verify this statement sufficiently well for present purposes by watching a vibrating string while it is coming to rest: as the amplitude of vibration diminishes so the sound gradually dies out. In the case of a tuning-fork it can be verified by a slight modification of Expt. 3.

But there are other circumstances that affect the intensity of the sound produced by an instrument. A vibrating string (or a tuning-fork) does not offer a large surface to the air: it cannot directly set any large mass of air into vibration, and consequently it produces only a feeble sound. But if it is made to impart its vibrations to an elastic body presenting a larger surface to the air (generally a thin wooden board called a 'sounding-board'), the intensity of the sound is greatly increased. This principle is employed in the construction of most stringed instruments, such as violins and pianos. Tuning-forks,

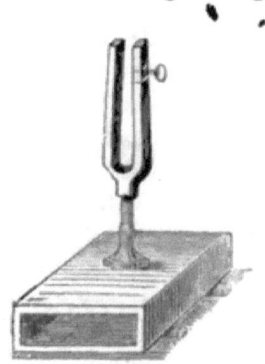

Fig. 5.

also, when used for experimental purposes, are generally mounted on sounding-boxes (Fig. 5), which greatly strengthen the sound.

EXPT. 4.—Take hold of a tuning-fork by the stem and set it into vibration. You hear only a feeble sound as long as the stem is held in the hand: but the sound swells out much more loudly when it is lightly pressed against a table or door-panel; or, better still, against its own sounding-box.

# CHAPTER II

## WAVE-MOTION

**11. Water-Waves.**—You cannot begin the study of wave-motion better than by examining carefully the waves which travel over the surface of a pool of water when a stone is dropped into the middle of it. Where the stone drops there is at first produced a depression, which immediately begins to spread outwards from the centre of disturbance in the form of a circular trough. The disturbance which thus travels to the edge of the pool is caused, not by any bodily motion of the water outwards, but by a downward motion of the water-particles, which spreads outwards from point to point. This downward motion is followed by a swing upwards, producing a crest, which follows the trough and travels after it with the same velocity across the surface of the pool. By this succession of moving troughs and crests is produced a series of ripples or water-waves. The distance between one crest and the next, or between one trough and the next, is called a *wave-length* (Arts. 15 and 16).

That the water itself does not move in the direction in which the waves travel is shown by the behaviour of chips of wood or bits of straw lying on its surface. These simply rise and fall, floating idly on the surface of the water and showing no tendency (unless the disturbance be very violent) to move forward in the direction in which the waves are travelling.

**12. Two kinds of Waves.**—Observe that in the case of water-waves the motion of the particles is an *up-and-down* motion; whereas the waves themselves move *horizontally* across the surface of the water. Such waves are called *transverse waves*, because they are produced by motion which is transverse or at right angles to the direction in which the waves are propagated. Strictly speaking, the motion of the water particles is circular, but all that we wish to do here is to point out the broad distinction between the two classes of waves, which are called respectively *transverse and longitudinal waves*.

**13. Transverse Waves** are such as are produced by a vibratory motion of particles executed in a direction at right angles to that in which the waves are propagated. As an instance of the production of a transverse wave may be mentioned the sudden jerk which a bargeman sends along a rope when he wishes it to clear an obstacle in its path.

Get a rope a few yards in length and lay it straight along the floor. Take hold of one end and, by jerking it rapidly from side to side, send a series of right- and left-handed pulses along it.

If an instantaneous photograph of the wave were taken it would present an appearance like that of the wave-line in Expt. 3. The waves appear to travel along the rope in the direction of its length: but the appearance is simply due to a vibration of each part of the rope which is executed transversely (or at right angles to its length) and which is communicated from each part of the rope to the next.

**14. Longitudinal Waves** are such as are produced by a vibratory motion of particles executed along the line in which the waves are propagated.

If you look down from a hillside upon a cornfield when a light summer breeze is blowing over it you will see a good example of wave-motion. The only motion of which the ears of corn are capable is a slight swinging motion; but as the breeze sweeps along the motion is transmitted from one ear to the next and from this again to its neighbour. Thus a certain state of things (ears of corn tightly packed together) is transmitted from point to point, and every gust of wind produces a wave of condensation which skims across the surface of the field. The waves are not wholly longitudinal. There is some little transverse motion: for each cornstalk swings like an inverted pendulum. But in the main the motion of the cars takes place in the line along which the wave travels. Observe that the characteristic of wave-motion is the transmission of a certain state of things or state of motion without any corresponding transmission of matter.

The best illustration of longitudinal waves is afforded by the behaviour of a spiral coil of wire.

EXPT. 5.—Make a spiral of thick copper wire by winding it tightly round a curtain-pole or a thick glass tube. When drawn out the spiral should be about two yards long.[1] Hang it up by a series of double threads, as shown in Fig. 6, the

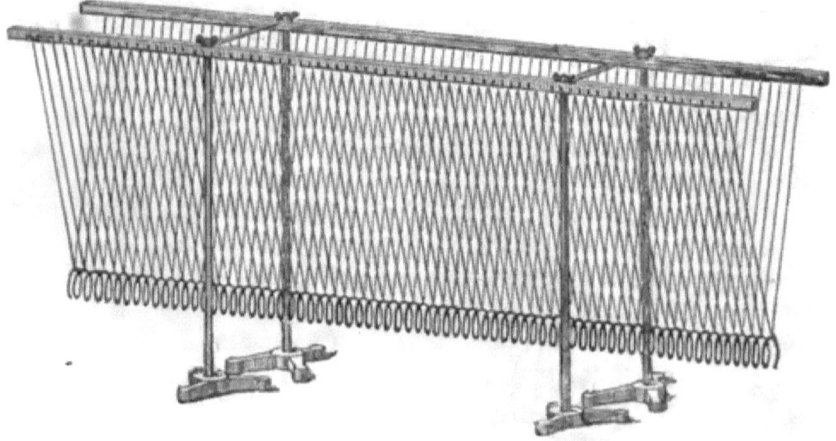

Fig 6.—WAVE-MACHINE (Weinhold).

double suspension being for the purpose of preventing any side swing.

By means of such a 'wave-machine' you can study the mode of propagation of waves of condensation as well as waves of rarefaction. A wave of rarefaction is produced by taking hold of one end of the spiral, pulling it out in the direction of the axis with a smart jerk, and then letting it go.

---

[1] A common defect of spiral coils used for illustrating wave-motion is that they are made too small and slight: when this is the case it is difficult to follow a wave as it quickly passes along the coil. It is really worth while taking trouble to make a good spiral, for it can be used afterwards to illustrate reflection, interference, and stationary vibration. I find that the following dimensions (recommended by Weinhold) are very suitable, viz.—Diameter of wire, 2 mm.; diameter of spiral, 7 cm.; number of turns, 72; length of completed spiral, 2 metres; length of threads 60 cm.

A wave of condensation is produced by striking one of the free ends of the spiral inwards. This is rather apt to make the spiral rock: a better plan is to take hold of the outer turn of wire by the end and push it inward or pull it outward, as the case may be.

Try both methods. Watch the waves as they run quickly along the spiral, and examine the way in which the separate coils move. You will thus learn more about wave-motion than you could by reading many pages of description. If you find it difficult to follow the motion of each turn of wire, gum strips of paper or tie bits of twine to a few of them: you will then easily see that each turn simply moves forward and backward as the wave passes it. Notice that when a pulse of condensation is running along the spiral the coils move *forward* in the direction in which the pulse is travelling: whereas in the case of a pulse of rarefaction the coils move *backward*, or in a direction opposite to that in which the pulse is travelling.

Place the blade of a knife between the coils near one end of the spiral and rake it quickly across a few turns towards the

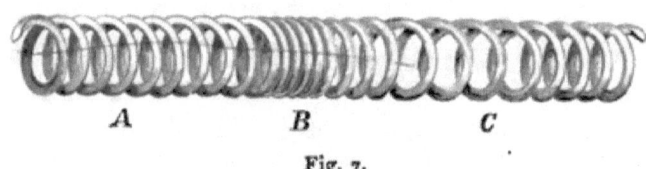

*A*    *B*    *C*

Fig. 7.

other end. You will thus produce a double pulse or complete wave: for each turn of wire as it escapes from the knife swings backward (or outward), thus producing a pulse of rarefaction (C, Fig. 7) which follows the pulse of condensation (B).

Fasten one end of the spiral to an empty box (a cigar-box). Strike the free end. The pulse runs along the spiral and you hear it deal a smart rap against the box. The energy of your blow is transmitted along the spiral to the box; the wave carries the energy without carrying the matter to which the energy was first imparted.

**15. Wave-line and Wave-length: Phase.**—The curved line in Fig. 8 represents a portion of a wave-line (or of a rope along which transverse waves are passing). The dotted line is the *axis* of the waves, and the arrow shows the direction in which they are travelling. The short equidistant vertical lines represent the paths along which the particles *a, b, c* ... vibrate.

The whole wave-line (of which only a portion is shown in the figure) consists of a repetition of exactly similar parts, and the shortest of such parts into which it can be cut is called a **wave-length**. Thus, starting from *a*, the distance *am* is a wave-length, for the part of the curve following *m* is an exact copy of that between *a* and *m*. Similarly *bn, co, dp* ... are wave-lengths.

The various particles *a, b, c* ... are moving with different velocities, some up and some down. *a, b, c* are moving downwards; *d* is at rest for an instant; from *e* to *i* all are rising: then *j* is at rest, and so on. Any two particles which are moving in the same direction and with the same velocity are said to be *in the same phase*. Thus the pairs *a* and *m*, *b* and *n*, *c* and *o*, *d* and *p* are in the same phase. We may therefore define the wave-length as being *the distance between the two nearest particles in*

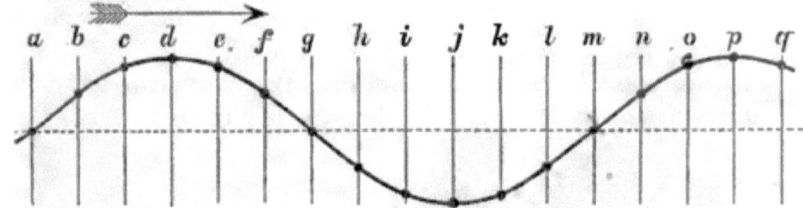

Fig. 8.—Transverse Wave-Motion.

*the same phase.* Notice particularly that *ag* is *not* a wave-length: it is only half a wave-length. *a* and *g* are not in the same phase: for although they have the same velocity they are moving in opposite directions. Nor are *d* and *j* in the same phase: for though both are at rest *d* is just starting downwards and *j* upwards.

**16. Wave-length and Velocity.**—The waves in Fig. 8 are supposed to be travelling towards the right. The curve indicates the positions of the particles at a particular instant. Their positions at any future instant can be found by shifting the wave-line forward through a certain distance depending on the interval of time.

Consider the particle *m*. When *m* has completed a whole vibration—down, up to the top, and back again to its mean position,—the disturbance will have moved forward through the distance *am*, *i.e.* through *a whole wave-length*. Thus we get a third definition of wave-length: it is *the distance through which a wave travels in the time required for a complete vibration.*

We know that

$$n = \frac{1}{\tau}$$

when *n* denotes the vibration-number and $\tau$ the period of vibration (Art. 3). Let $\lambda$ denote the wave-length and *v* the velocity with which the waves travel forward : we have just seen that they move through the distance λ in the time τ.

Thus since $\text{velocity} = \dfrac{\text{distance}}{\text{time}}$,

it follows that $v = \dfrac{\lambda}{\tau} = \dfrac{1}{\tau}.\lambda$,

or $v = n\lambda$.

This last result is of the greatest importance. The relation between velocity, vibration-number, and wave-length which it expresses holds good not only for transverse waves but for *all* kinds of waves.

**17. Comparison between Longitudinal and Transverse Waves.**—The bottom line (R) in the accompanying figure represents a series of particles at rest. The line above (L) represents the positions of the particles when traversed by a series of longitudinal waves. The top line (T) represents the positions of the particles at the same instant when traversed by corresponding transverse waves. An upward displacement in the transverse wave corresponds to a displacement to the right in the longitudinal wave : a downward displacement in the transverse wave corresponds

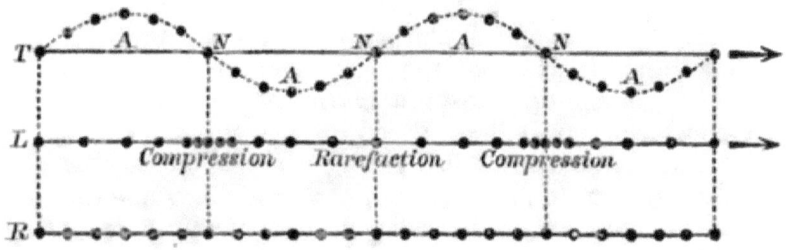

Fig. 9.—Transverse and Longitudinal Waves.

to a displacement to the left in the longitudinal wave. The position of any particle in L is found by displacing the corresponding particle in R through a distance proportional to the displacement in T. The two sets of waves correspond in the following respects :—

A *crest* in T corresponds to portions of L in which particles are displaced to the *right*. A *trough* corresponds to displacements to the left.

The points of no displacement (N) occur alternately on falling and rising parts of the curve T : these correspond to alternate centres of compression and rarefaction in L.

In T the distance NN represents *half* a wave-length : so in L the distance between a pulse of compression and a pulse of rarefaction is *half* a wave-length. A *complete* wave-length in L is represented by the distance between two successive condensations or two successive rarefactions.

# CHAPTER III

## TRANSMISSION OF SOUND—ITS VELOCITY

**18. Transmission of Sound through Air.**—We have seen that sound is caused by a vibratory motion of the sounding body. We have next to consider how this motion travels through the air and reaches our ears.

It is important to understand that when sound is transmitted through the air, the air itself does not move as a whole, but a certain state of motion is transmitted through the air from point to point and from particle to particle. Thus the report caused by the explosion of a gun can be heard for miles around; but the smoke expelled from the muzzle of the gun only travels forward for a few yards.

Let us return to our spring (Fig. 1) and see how its vibrations affect the air in its neighbourhood. As the spring swings from left to right ($l$ to $l'$) it compresses the air in front of it. The air, being elastic, resists the compression and tends to expand. In so doing it compresses the air lying in front of it. This again reacts upon the next layer of air, and so the motion of the spring causes a *pulse of compression*, which travels forward with uniform speed through the air.

Meanwhile the spring has swung backwards towards the left hand (from $l'$ to $l$). The particles of air in contact with it share its motion, and thus the layer of air lying towards the right of the spring expands or is rarefied. The pulse of rarefaction thus produced travels forward through the air with the same uniform speed as the pulse of compression.

As the spring begins its next vibration another pulse of compression is produced; and this again is followed by a pulse of rarefaction. Thus a series of alternate pulses of compression and rarefaction are produced, and follow each other in regular succession through the air (as indicated in Fig. 10). The student must refer to text-books on Physiology for an explanation of the manner in which these sound-waves affect the ear when they fall upon it and produce the sensation of sound.

The wave-length of a given note in air may be defined in any of the ways given in Arts. 16 and 17: *e.g.* it is the

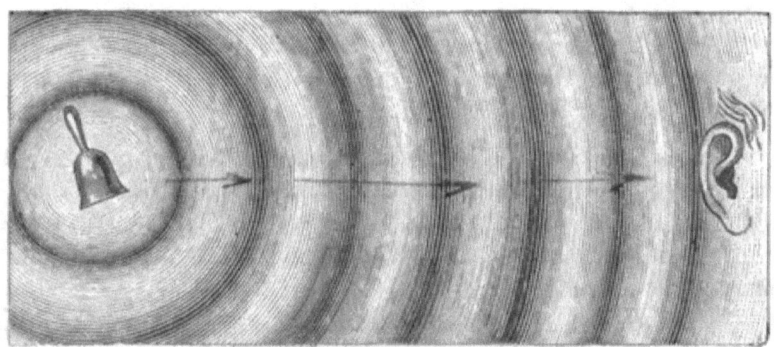

Fig. 10.—Sound-Waves in Air.

distance between two consecutive pulses of compression or two consecutive pulses of rarefaction; or it is twice the distance between a pulse of compression and the next pulse of rarefaction.

**19. Transmission through a Cylindrical Tube.**—Sound-waves are propagated through air contained in a cylindrical tube much like the waves in Expt. 5. The waves are longitudinal: the air-particles vibrate backward and forward in the line along which the waves advance.

Imagine the spiral replaced by a cylindrical column of air, along which pulses are sent by a vibrating tuning-fork or spring at one end. Each forward swing of the fork sends a pulse of condensation along the column: the ensuing backward swing sends a pulse of rarefaction after it at the same speed. A regular succession of such pulses constitutes a train of sound-waves. This gives you a picture of how sound travels along air confined in a hollow tube.

**20. Intensity.**—In general, however, the disturbance travels outwards in all directions from the source. All the particles which are just beginning to move at any instant are said to *lie in the wave-front*. In free air the wave-front has the form of a sphere.

In calm air the sound produced by a bell or other vibrating body is heard equally well in all directions around it. The sound-waves are spherical in form (the source of the sound being the centre of the spheres), and they travel outwards freely in all directions.

It can be shown from this that the intensity of the sound diminishes according to the same law (the 'law of inverse squares') as the intensity of light (p. 124).

**21. A Material Medium required for Transmission of Sound.**—Before proceeding further, it will be well to point out that sound cannot, like light and radiant heat, travel through a vacuum. This may be shown by the following experiment.

EXPT. 6.—Place on the plate of an air-pump a few folds of flannel or a thick tuft of cotton-wool. On this lay a loud-ticking watch, an alarum clock, or any arrangement for striking a bell by clockwork (Fig. 11). Cover the whole with a bell-shaped receiver, and begin working the pump. As the air is gradually exhausted from the receiver the sound becomes fainter and fainter. If the pump works well and produces a good vacuum, the sound may entirely disappear. The effect is still more strikingly shown by readmitting the air into the receiver, when the sound is again distinctly heard.

Fig. 11.

Air is not the only gas that will transmit sound. This may be shown by admitting any other gas into the exhausted receiver.

**22. Sound Transmitted by Liquids.**—It is said that divers under water can distinctly hear words spoken by persons near the shore. That water does conduct sound may be easily shown as follows.

EXPT. 7.—Stick the handle of a tuning-fork into a large cork. Take a tumbler, nearly fill it with water, and place it on a tray or on the sound-box belonging to the fork. Strike the fork, wait until the sound has become faint, and then hold it so that the cork is immersed in the water. There is a marked increase in the loudness of the sound. This is more strikingly shown by alternately raising the cork and again lowering it into the water, when the intermittent swelling and sinking of the sound is plainly perceived. The explanation will be easily seen on referring to Art. 10. The sound is conducted through the water to the glass and the sounding-board, which in turn sets the air into vibration.

**23. Sound Transmitted by Solids.**—The following experiments will serve to illustrate the transmission of sounds by solid bodies.

EXPT. 8.—To one end of a pine rod (4 or 5 ft. long and about 1 in. in diameter) glue a thin pine board about 6 in. square to act as a sounding-board. To the other end hang a watch by a hook, and then apply your ear to the sound-board. Notice how distinctly the ticking of the watch is heard.

Or get a friend to strike a tuning-fork and press it against the farther end of the rod, moving it off and on so that you may be certain that the sound is heard by transmission through the rod.

EXPT. 9.—Stop up both your ears and get a friend to hold a watch so that you can take hold of the ring between your teeth. Do this and you will be surprised at the loudness of the ticking. The sound seems to be felt rather than heard. Bite off and on to make sure that it does not come through the air. The sound is conducted through the teeth to the bones of the head, and through them to the ears.

**24. Velocity of Sound in Air.**—Just as waves travel with a definite velocity across the surface of water, or along spiral springs, so are sound-waves propagated with a definite velocity through air. This velocity is vastly smaller than that at which radiant heat and light travel (p. 131). Thus we see lightning-flashes some time before we hear the thunder that accompanies them, and the interval that elapses between the flash and the peal gives us some idea how far off the storm is.

The velocity of sound in air has been measured repeatedly and by various methods. Of these the most easily understood is that adopted by the Paris Academicians. This consisted in measuring the interval that elapsed between the instant at which the flash from a distant cannon was seen and the instant at which the report was heard (the exceedingly short time taken by light in travelling over the same distance being, of course, neglected). The same method was employed afterwards by Arago, Gay-Lussac, and others. They chose two stations (Villejuif and Montlhéry) near Paris and 18,613 metres apart. The observers at each station were furnished with chronometers and a 6-pounder cannon. Observations were made at both stations alternately so as to get rid of any disturbing effect due to wind. As the mean result of many experiments it was found that the report took 54.6 seconds to travel from one station to the other. Dividing the distance by the time, we get 341 metres as the distance travelled in one second. Allowing for the temperature (16° C.) at which the experiment was made, this gives 331 metres per second as the velocity of sound in air at 0° C. Subsequent experiments have shown that this value is correct. It corresponds to a velocity of 1086 feet per second.

25. The velocity of sound in air *increases with rise of temperature* at the rate of about 2 ft. or 62 cm. per degree centigrade. Thus at the ordinary temperature of 15° sound travels about 1116 ft. or 340 metres per second.

The velocity is **independent of the** atmospheric pressure. However the barometric height may vary, sound travels at the same rate as long as the temperature remains constant (see Arts. 30 and 31).

The velocity is practically independent of the loudness of the sound and *quite independent of pitch*. Otherwise a song would only be heard in correct time close to the singer, and harmony would be impossible; band-music heard a hundred yards away from the band would consist of most frightful discords.

**26. Velocity in Water.**—The velocity of sound in water was measured in 1827 by Colladon and Sturm. They moored two boats in the Lake of Geneva at a distance of 13,487 metres apart. From one boat a large bell B (Fig. 12) was suspended in the water. This could be struck by a hammer $h$ attached to a lever which was so arranged that in moving it a match $m$ was brought into contact with a heap of powder $p$ at

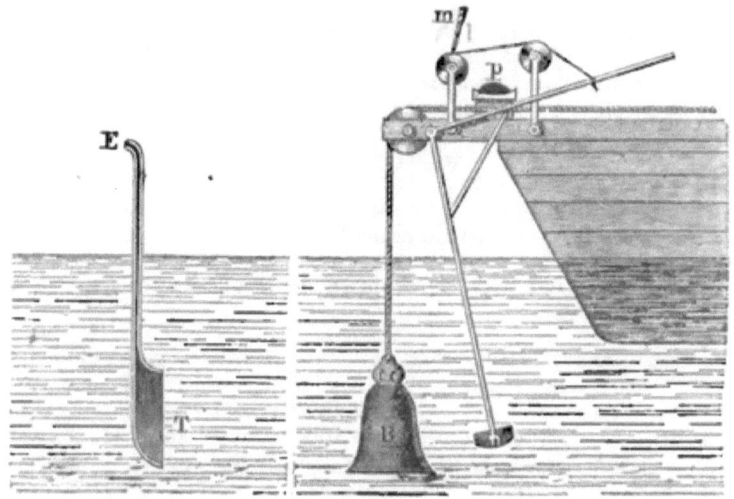

Fig. 12.

the very instant when the bell was struck. The flash served as a signal to the observers in the second boat. From this was suspended in the water a kind of large speaking-trumpet, the mouth of which was closed by a membrane T. By applying his ear at E, the observer could distinctly hear the sound transmitted through the water. It was found that the interval between seeing the flash and hearing the sound was 9.4 seconds. This gives 1435 metres per second as the velocity of sound in water.

**27. Velocity in Solids.**—Sound travels very rapidly through solids, its velocity in glass and steel, for example, being about fifteen times as great as its velocity in air.

Sound is also transmitted through rocks, and with greater velocity than in air. If you stand near a mine in which shot-firing is going on, you can generally distinguish two sounds, the first being a dull thud or rumbling noise travelling through the ground, and the second being the report transmitted in the ordinary way through the air.

**28. Theoretical Calculation of the Velocity in Air.**—According to a theoretical calculation made by Newton, sound ought to travel in any medium with a velocity which is given by the equation

$$V = \sqrt{\frac{E}{D}} \quad \quad \quad \quad \quad (1)$$

when E denotes the elasticity of the medium and D its density.

Now it can be proved that in the case of air or any other gas kept at a constant temperature *the elasticity is equal to the pressure*. Thus the velocity in air should be given by the equation

$$V = \sqrt{\frac{P}{D}} \quad \quad \quad \quad \quad (2)$$

when P denotes the atmospheric pressure, expressed in the proper units.

Inserting the correct numerical values for P and D in the above equation, it is found to give about 280 metres or 920 feet per second as the velocity of sound in air at 0°. This is nearly 16 per cent less than the velocity found by experiment.

**29. Laplace's Correction.**—Evidently there was something wrong about the calculation. Newton tried to explain how the difference arose, but the true explanation was given afterwards by Laplace. He pointed out that the compression produced by a sound-wave takes place so rapidly that the heat which is developed by it remains in the wave of compression: there is not sufficient time for it to be lost by conduction or radiation. Again, each wave of rarefaction produces a cooling effect (p. 109); and thus the elasticity cannot be calculated on the assumption that the temperature remains constant. The effect of all this is to increase the elasticity and make it equal to about 1·41 times the pressure. Introducing this correction (known as Laplace's correction) into the equation we get

$$V = \sqrt{\frac{1 \cdot 41\, P}{D}} \quad \quad \quad \quad (3)$$

and this gives a value which agrees almost exactly with the velocity found by experiment.

**30. Effect of change of Pressure.**—The velocity of sound in air is not affected by any change in the atmospheric pressure (or barometric height). This may be seen by considering how a change of pressure affects the two factors E and D which occur in the expression for the velocity. Now it is true that the elasticity is proportional to the pressure and increases as the pressure increases. But it follows directly from Boyle's Law that the density of any gas is also directly proportional to the pressure. Any increase of pressure increases both E and D *in the same proportion*, and hence their ratio $\left(\frac{E}{D}\right)$ remains constant. Thus the velocity is independent of the pressure.

**31. Effect of Temperature.**—It is otherwise when the temperature of the air changes. For suppose that the temperature rises and consider how this will affect the quantities in equations (1) or (3). Since the air is free to expand, the pressure or elasticity will remain unaltered. But the density of the air diminishes just as its volume increases (the density of a given mass of gas being inversely proportional to its volume). Thus rise of temperature produces an increase in the velocity, as stated in Art. 19.

If the temperature rises from $0°$ to $t°$, the volume increases in the ratio of $1$ to $(1+at)$, and the density diminishes in the ratio of $(1+at)$ to $1$. From this it may easily be seen that if $V_0$ denote the velocity of sound in air at $0°$, its velocity at any temperature $t°$ will be

$$V_t = V_0 \sqrt{1+at}.$$

**32. Velocity in other Gases.**—The equations for the velocity given in Arts. 28 and 29 hold good not only for air but also for other gases. It therefore follows that the velocity in any gas is inversely proportional to the square root of the density of the gas.

For example, oxygen is sixteen times as dense as hydrogen. Hence the velocity of sound in oxygen is to that in hydrogen as $\sqrt{1} : \sqrt{16}$, or as $1 : 4$.

# CHAPTER IV

## REFLECTION OF SOUND

**33. Reflection of Longitudinal Waves.**—A good idea of the way in which sound-waves are reflected may be obtained by studying the reflection of longitudinal waves at the end of a wire spiral such as was used in Expt. 5. The reflection takes place in two distinct ways, according as the end is *free* or *fixed*. In both cases we shall suppose the original (or incident) pulse to be a pulse of compression.

EXPT. 10. **Reflection from a Free End.**—The spiral is supposed to be in the state shown in Fig. 6 (both ends free). A pulse of compression is sent from one end—say the right-hand: when it reaches the other end, it does not disappear but is reflected *as a pulse of rarefaction.* When it returns to the right-hand end it is again reflected, but now as a pulse of compression again. Thus at each reflection at a free end the wave suffers a *change of type*: a pulse of rarefaction being reflected as a pulse of compression and *vice versâ*.

EXPT. 11. **Reflection at a Fixed End.**—Push one end of the spiral into a cork and clamp this fast to a heavy retort-stand. Send a pulse of compression along the spiral from the free end. When the pulse reaches the fixed end the motion of the coils is reversed and the wave is reflected back, but still as a pulse of compression. In the same way a pulse of rarefaction is reflected as a pulse of rarefaction. In reflection from a fixed end there is *no change of type.* The cases of reflection to which we shall for the present restrict our attention are of the same kind as those which occur at the *fixed* end of a spiral.

**34. Reflection of Sound-Waves.**—Sound-waves are reflected from solid obstacles in much the same way as the longitudinal waves in the last experiment. Each pulse of compression or rarefaction, when it reaches the obstacle, is reflected, giving rise to a pulse of compression or rarefaction travelling in the opposite direction.

In the case of light we saw that in order to get good reflection it was necessary that the reflecting surface should be smooth and polished : in the case of sound the most essential thing is that the reflecting surface should be of considerable extent. You will doubtless recollect instances in which you have heard band-music or the sound of church-bells apparently reaching you in a direction totally different from that of the band or steeple, the sound being reflected from a wall or row of houses.

**35. Echoes.**—An echo is produced when a sound is reflected normally (or back along its own path) from a high cliff or wall. If a person standing at a sufficient distance from the cliff shouts or claps his hands, he hears the sound repeated. In order that he may hear the echo distinctly it is necessary that the time taken by the sound in travelling to the cliff and back again should be sufficiently long to enable him to distinguish between the original and the reflected sound, for otherwise the two would be confused together.

Suppose that we allow one-fifth of a second for this ; and further that we take the velocity of sound as 1100 ft. per second. In one-fifth of a second sound travels 220 ft., so that we must allow for 110 ft. to the cliff and 110 back to the observer. Thus in order to hear the echo distinctly the observer's distance from the cliff should not be much less than 110 ft.

**36. Sensitive Flames.**—In experimenting on reflection of sound (and especially for class demonstration) it is convenient to have some means of indicating the presence and whereabouts of a sound. This is best done by means of sensitive flames.

When the pressure of the gas supplied to a jet is gradually increased, a point is reached at which the flame begins to flare: just at this point it becomes very sensitive to sound and especially to hissing and tinkling sounds. The flame should be long, thin, and straight, as shown in the accompanying figure. A suitable jet can be made by drawing out a piece of glass tubing to about 2 mm. diameter, cutting it off with a file and grinding the edge smooth on a stone. But as glass jets are very liable to crack, it is better to get a plain steatite burner with a small round hole. The sensitiveness of a flame depends upon the burner and upon a proper regulation of the gas-pressure: sometimes a burner is so sensitive that it is scarcely possible to work with it. The flame can be protected from accidental noises by surrounding it with a glass globe provided with suitable openings; into one of these may be inserted a funnel for the purpose of catching the sound and directing it towards the base of the flame.

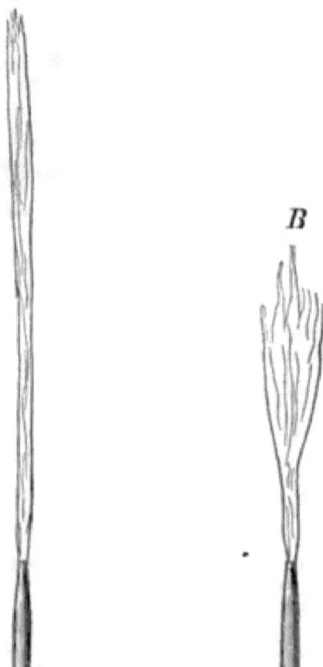

Fig. 13.

EXPT. 12.—Procure a gas-bag (a rubber bag such as is used for lantern work), press it down flat to squeeze out all air, and then connect it by rubber tubing to the gas-supply. Raise the pressure-boards, so as to let the gas stream in freely, and when the bag is full of coal-gas close the tap attached to the bag. Connect the tubing to the sensitive jet. Put on the pressure-boards and place weights on them so as to increase the pressure until the jet begins to flare. Now reduce the pressure slightly and you will find that the flame is in the sensitive state. It has the form shown in A, Fig. 13. The slightest noise makes the flame roar, and at the same time it becomes shorter and broader and jagged at the edges, as shown in B.

Stand a few feet from the flame and shout : each time it ducks down. Try singing and whistling. Talk to it and see how it picks out the *s* sounds. Rattle a bunch of keys : it roars furiously.

**37. Experiments on Reflection of Sound.**—The reflection of sound may be illustrated in the same way and with the same apparatus as that described on pp. 99-101 for demonstrating the reflection of heat-radiation.

EXPT. 13.—Arrange the tin tubes and reflector as shown in Fig. 59, p. 100. For the present purpose the tubes may be somewhat longer than those used for the radiation experiment : say 4 ft. long each. At T place the sensitive flame, adjusting the mouth of the funnel over the end of the tube. It may be necessary to screen the flame from sound travelling directly from B to it : this can be done by hanging wet towels across between B and T to act as sound-screens. Or the sound may be prevented from getting outside as follows : Make a loose sleeve of cloth and tie it over the end of the tube at B. Take a small piece of tinp-late in one hand and a penny in the other. Place both hands well inside the sleeve and tap the penny against the tin-plate. If the reflector at R is properly adjusted, the flame will now respond, ducking down at each tap. Remove the reflector R. If the flame is still affected, try tapping more gently.

EXPT. 14.—Repeat the experiment by hanging a watch in front of the tube at B and using your ear in place of the sensitive flame at T. The ticking is much more distinctly heard when the reflector is in position than when it is removed.

EXPT. 15.—Place the two tubes end to end and in a straight line so as to form one tube 8 ft. long. At this distance the ticking of a watch is not audible in free air. But if you hang a watch in front of one end of the tube and apply your ear to the other end, you will hear the ticking with surprising distinctness.

This will enable you to understand how 'speaking-tubes' can be used for communicating between different rooms in a building. Their behaviour may be explained as a result of repeated reflections from the sides of the tube (see **Expt. 64, p. 100**), or as **a** result of keeping the disturbance **constantly** within the walls **of the tube.**

### Examples on Chapters I-IV.

1. **State** exactly what you **mean by the** '**density**' **of a** medium, and **explain** what you understand by its 'elasticity.' Is the elasticity of water greater or less than that of air? How would you prove to any one that your answer is correct?

2. Describe and explain **an** experiment illustrating the use **of sound-**boards in making the vibrations of a wire audible.

3. Explain any method **by means of which the** ticking of **a watch may** be made audible to a person **at the other end of a** large room.

4. What is the velocity **of sound in air?** On what grounds **can it be** asserted that musical tones of high **and low pitch** travel at the same speed?

5. What is meant by a *wave of sound* and by the *length of a wave*. Explain how sound is transmitted through air.

6. A **gun is** fired **on** a cold winter's day at a certain distance from an observer who hears the report five seconds after seeing the flash. Would the interval between seeing the flash and hearing **the** sound have been the same **on** a hot day in summer? **Give** reasons for your answer.

7. **A street** with houses on **both sides** runs north and south, and a church **is** situated at a little distance **to the** east of it. **As** I walk down the eastern **side** of the street the sound **from a peal of bells** in the church-tower seems **to** come from **the** west. **Explain this,** drawing a diagram to illustrate **your** answer.

# CHAPTER V

## PITCH AND MUSICAL INTERVALS

**38.** Musical sounds may differ in respect of—
    (1) **Intensity or Loudness.**
    (2) **Pitch.**
    (3) **Quality.** or Timbre

We have already stated that the intensity or loudness of a sound depends upon amplitude of vibration (Art. 10). Pitch we shall consider more fully in the present chapter. At this stage it is not possible to treat of quality of sounds. What is meant by difference in quality may be readily appreciated on producing the same note by means of different instruments, say a piano, a violin a flute and the human voice. Even though the sounds be of exactly the same pitch and, as nearly as possible, of the same loudness, you can easily tell which is which. As we go on to consider the different ways in which musical sounds are produced we may be able to point out some of the causes which produce these differences in quality.

**39. Pitch** is that which distinguishes a high or shrill note from a deep or low one. Most people can tell when two notes are of the same pitch or height; and can decide which is the higher of two notes. People who are able to judge correctly of differences in pitch are said to have 'a good ear for music.' If you are blessed with such, you will already know what is meant by pitch; and if you are not, it is scarcely possible to help you to such knowledge.

**40.** We proceed to show by experiment (1) that any series of regularly recurring impulses (within a certain range) produces a musical note; and (2) that the more rapidly these impulses follow each other, the higher is the pitch of the note.

EXPT. 16. **Savart's Wheel.**—On a centrifugal machine or whirling-table (see Fig. 15) fix a toothed wheel, such as a large clock-wheel. Make the wheel rotate, at first very slowly, then more and more rapidly, at the same time holding a thin card lightly against the teeth. At first you hear only a series of separate taps. As these succeed each other more rapidly they begin to blend together and produce a low note the pitch of which increases as the rapidity of rotation increases. The quality of the sound produced is poor and thin: it is scarcely correct to call it a musical note.

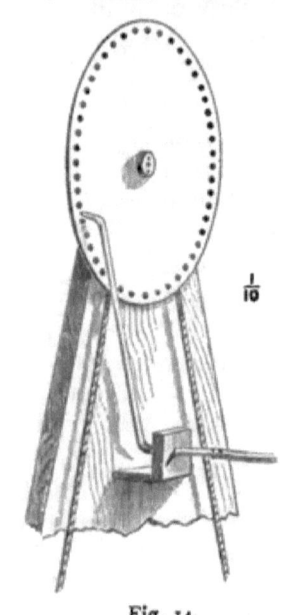

Fig. 14.

EXPT. 17. **The Disc-siren.**—Perform a similar experiment by blowing a jet of air against a rotating disc having a circular row of holes pierced in it (Fig. 14). Here again the quality of the sound is poor: it is mixed up with the noise made by the air in rushing against the disc.

The disc may be made of stiff smooth cardboard or sheet-metal and of about the following dimensions. Radius of disc, 15 cm. Radius of row of holes, 13 cm. Holes about 0.5 cm. in diameter and 2 cm. apart. Jet made of glass-tubing drawn out at the point to same diameter as the holes or somewhat narrower.

What is properly known as a 'siren' is an instrument which does not differ greatly in principle from the above. It is, however, more elaborate in construction and contains a 'counter,' which registers the number of revolutions made in a given time. When this is known we can easily calculate the number of puffs of air per second that are required to produce a given note.

In neither of the above experiments is the sound produced directly by a vibratory motion. In the case of the disc-siren, each time a hole comes in front of the jet the air rushes through and produces a pulse of compression in the space beyond. In virtue of the elasticity of the air, this is succeeded by a pulse of rarefaction during the interval that elapses before the next hole comes into position. Thus the air is set into vibration much as it would be by the motion of a tuning-fork or vibrating string. The pitch of the note produced increases as the vibration-number or frequency increases (Art. 3).

**41. How Pitch is expressed.**—We may express the pitch of a note—

(1) **Relatively**, as when we say that one note is an octave or a fifth higher than some other note which is chosen as a convenient standard of reference. This is the method generally adopted in Music.

(2) **Absolutely**, as when we say that a certain number of vibrations per second is required to produce a note of a given pitch. This is the method adopted in Physics; the pitch of any note being expressed by stating its vibration-number or frequency (Art. 3). Thus the pitch of the note produced by an open organ-pipe 2 feet long is 280; for such a pipe, at the ordinary temperature of the air, produces vibrations at the rate of 280 per second.

**42. Musical Intervals.**—The interval between any two notes is measured physically by the ratio between the vibration-number of the higher note and that of the lower one. If the vibration-number of the lower note be $n$, and that of the higher note $n'$, the interval is measured by the ratio $\frac{n'}{n}$.

*Unison.*—Two notes are said to be in unison when they have exactly the same pitch. The ratio between the vibration-numbers of two such notes is clearly $\frac{1}{1}$. Thus although there is no difference of pitch between two notes in unison, the interval is expressed by 1 and not by 0.

*Octave.*—If the vibration-number of one note is double that of another, the first note is said to be an octave above the second. This interval is represented by the ratio $\frac{2}{1}$. Like all other musical intervals, its value depends, not on the absolute vibration-numbers, but on their ratios. The notes represented by the numbers 100, 200, 400, 800 . . . are each an octave above the one below.

**43. The Major Diatonic Scale.**—Musicians divide the interval between a note and its octave into seven smaller intervals of unequal value known as tones and semi-tones.

The notes which occur in this 'scale' may be typified (although they are not exactly represented) by the white keys of a pianoforte. In the old or 'staff' system of notation these notes are known as

    C    D    E    F    G    A    B    c.

In the new or 'Tonic Solfa' system of notation (for which we shall always use thick letters), the notes are represented as

    **d**    **r**    **m**    f    **s**    **l**    **t**    **d'**

which are read as

    Doh    Ray    Me    Fah    Soh    Lah    Te    doh.

These seven notes (or eight, with the addition of the octave[1]) form the musical scale in common use. To distinguish it from a somewhat different scale, used chiefly in solemn and mournful music (the minor scale, called by Solfaists the 'lah mode'), it is known as the *major* diatonic scale.

At this stage no better advice can be given to a non-musical student than that he should play these notes on the piano, or, better still, get some one to sing them to him, until he knows the different notes and the intervals between them. Without such knowledge any discussion of the nature of musical intervals must be as unintelligible as Chinese. He should also listen carefully to the notes when sounded in pairs, and notice what combinations of the notes are harmonious (or produce a pleasing effect upon the ear) and which are dissonant (or produce a harsh or disagreeable effect).

**44. Intervals which occur in the Common Chord.**—We now proceed to illustrate how some of the more important intervals that occur in music can be measured and expressed by numbers. For this purpose we shall choose the most harmonious intervals, viz. those which occur in the Common or Major Chord. The notes which form this chord are

            **d**      **m**      **s**      **d'**

or,      C      E      G      c.

We shall show in two ways that in order to produce this series of notes the vibration-numbers must be in the proportion of

    4,    5,    6,    8.

**1. By Savart's Wheels**—EXPT. 18.—Four toothed wheels will be required, having respectively 80, 100, 120, and

---

[1] Latin *octavus* = eighth.

160 teeth (or numbers in the proportion of 4, 5, 6, and 8); and the wheels are to be fixed (at a little distance apart) on the axis of a whirling-table, as shown in Fig. 15.

Rotate the wheels and touch each in succession with a thin card as in Expt. 16. As the speed of rotation cannot be kept constant for any length of time it is best to touch the wheels very lightly and rapidly one after the other; the notes are easily recognised as being those which form the Common Chord.

Change the rate of rotation. The notes all alter; but your ear perceives that they always bear to each other the same 'relative pitch.' The intervals do not depend upon absolute pitch, but only upon the *ratios* of the vibration-numbers.

II. **By the Disc-siren**—EXPT. 19.—For this purpose we require a disc of the same size as that used in Expt. 17, but pierced with four circular rows of holes (Fig. 15). The innermost row should have 24 holes, the next 30, the next 36, and the outer row 48 (the numbers being in the proportion of 4, 5, 6, and 8).

The jet may be made of glass tubing bent twice at right angles and mounted as shown in Fig. 14: this arrangement is convenient for swinging the jet quickly over the four rows of holes without fear of breaking it.

**45. How Intervals are Compounded.**—Musical intervals are compounded *by multiplication* and not by addition.

The experiments just performed show us that the interval between **d** and **m**, or C and E, is represented by any of the equal ratios

$$\frac{100}{80}, \quad \frac{30}{24}, \quad \text{or} \quad \frac{5}{4}.$$

Fig. 15.

The interval between **m** and **s**, or F and G, is represented by any of the equal ratios

$$\frac{120}{100}, \quad \frac{36}{30}, \quad \text{or} \quad \frac{6}{5}.$$

Again the interval between d and s, or C and G, is represented by

$$\frac{120}{80}, \quad \frac{36}{24}, \quad \text{or} \quad \frac{3}{2}.$$

Now this last is also the interval obtained by compounding the first two, for

$$\frac{5}{4} \times \frac{6}{5} = \frac{6}{4} = \frac{3}{2}.$$

**46. Intervals which occur in the Scale.**—By experiments similar to those above described, or by careful experiments made with more elaborate apparatus, the ratios of the vibration-numbers of the notes forming the scale have been determined. The names and values of the various intervals, counting from C (the tonic or key-note) are given in the following table. The first column indicates the notes numbered 1, 2, 3 ... in order, and the second and third columns their names. The fourth column gives the names of the intervals, and the fifth their numerical values.

| | | | | |
|---|---|---|---|---|
| 1 : 2 | d : r | C : D | Second | $\frac{9}{8}$ |
| 1 : 3 | d : m | C : E | (Major) Third | $\frac{5}{4}$ |
| 1 : 4 | d : f | C : F | Fourth | $\frac{4}{3}$ |
| 1 : 5 | d : s | C : G | Fifth | $\frac{3}{2}$ |
| 1 : 6 | d : l | C : A | (Major) Sixth | $\frac{5}{3}$ |
| 1 : 7 | d : t | C : B | (Major) Seventh | $\frac{15}{8}$ |
| 1 : 8 | d : d¹ | C : c | Eighth or Octave | $\frac{2}{1}$ |

Of these intervals the Second $\left(\frac{9}{8}\right)$ and the Seventh $\left(\frac{15}{8}\right)$ are dissonant. The others are consonant or harmonious, the most perfect consonance being given by the Fifth $\left(\frac{3}{2}\right)$ and the Octave $\left(\frac{2}{1}\right)$. Thus we see that the most harmonious intervals are those in which small numbers (from 1 to 5) occur.

**47. Standards of Pitch.**—Once we settle upon the absolute pitch of the tonic or key-note of our scale, the absolute pitch of every other note in the scale is thereby fixed. Thus if we agree to take C=256, the vibration-number of G (a fifth above it) is $256 \times \frac{3}{2} = 384$, and so on. This is the pitch usually adopted by writers on acoustics and by makers of acoustical apparatus. The number 256 has the advantage of being a power of 2 (viz. $2^8$), and the choice is mainly a matter of convenience.

The most convenient standard of pitch is a tuning-fork made so as to execute a known number of vibrations per second. In our country there is no legal standard of pitch. What is vaguely referred to as 'concert pitch' may be taken to represent C=264, or thereabouts. (The C here referred to is what is known as the 'middle C' of a piano.)

48. **Intervals between successive Notes in the Scale.**—The various intervals given in Art. 46 have as their least common denominator the number 24. If we take this as representing the vibration-number of the key-note, the other notes in the scale will be represented by the following whole numbers:—

| 24 | 27 | 30 | 32 | 36 | 40 | 45 | 48 |
|----|----|----|----|----|----|----|----|
| d  | r  | m  | f  | s  | l  | t  | d$^1$ |
| C  | D  | E  | F  | G  | A  | B  | c  |
| $\frac{9}{8}$ | $\frac{10}{9}$ | $\frac{16}{15}$ | $\frac{9}{8}$ | $\frac{10}{9}$ | $\frac{9}{8}$ | $\frac{16}{15}$ | |

The numbers in the last row represent the intervals between each successive pair of notes. They are obtained by dividing each number in the top row by the one before it. Thus the interval r : m is equal to $\frac{30}{27}$ or $\frac{10}{9}$. This is not quite the same as the preceding interval $\left(\frac{9}{8}\right)$, but both are known as *tones*. The interval $\frac{16}{15}$ is called a *semi-tone*. Thus the major diatonic scale consists of a series of intervals in the following order: two tones and a semi-tone, three tones and a semi-tone.

49. **A Harmonic Series is a series** of notes whose vibration-numbers are in the following proportion—

$$1 \quad 2 \quad 3 \quad 4 \quad 5 \quad 6 \ldots$$

All the notes in such a series (at any rate up to the sixth) harmonise well with the first (or fundamental) note and with each other. It will be a useful exercise to find out the relations between these notes.

Let us call the lowest or fundamental note d. The interval between this and the next note is an octave $\left(\frac{2}{1}\right)$: hence the second note is d$^1$. The third note is a fifth $\left(\frac{3}{2}\right)$ above the second, or a twelfth above the first; it is therefore the note s$^1$. The fourth note is an octave $\left(\frac{4}{2}=\frac{2}{1}\right)$ above the second, or two octaves $\left(\frac{4}{1}\right)$ above the first; it is therefore d$^{11}$. The fifth is a major third $\left(\frac{5}{4}\right)$ above the fourth, and is therefore $m^{11}$. The sixth is a fifth $\left(\frac{6}{4}=\frac{3}{2}\right)$ above number 4, and is therefore $s^{11}$. Thus the first six notes of the harmonic series are

| 1 | 2 | 3 | 4 | 5 | 6 |
|---|---|---|---|---|---|
| d | $d^1$ | $s^1$ | $d^{11}$ | $m^{11}$ | $s^{11}$ |
| C | c | g | $c^1$ | $e^1$ | $g^1$. |

# CHAPTER VI

## TRANSVERSE VIBRATIONS OF STRINGS

**50.** By a string is here meant any elastic and flexible cord (such as the twisted cat-gut used in violins) or metallic wire stretched between fixed supports. When such a stretched string is pulled to one side it tends to return to its position of rest. When let go it flies back, but, like a spring or pendulum, it overshoots the mark, and goes on swinging from side to side. A string can be set into vibration by striking, plucking, or bowing it. The vibrations thus produced are *transverse vibrations*, and these are the only ones that we shall consider. The vibrations are further said to be *stationary*: certain points (*e.g.* the fixed ends) remain permanently at rest. These points are called *nodes*.

We shall see presently that a string may vibrate transversely in many different ways, dividing up into a number of smaller vibrating parts or segments. The lowest or *fundamental* note of the string is produced when it vibrates as a whole (or in one segment). In this case the only nodes are at the two fixed ends. All other points are in motion, and the amplitude of the motion is greatest at the centre, which is called an *antinode*.

**51. Laws.**—The number of vibrations executed per second by a string when sounding its fundamental note is found to depend upon

(1) Its length.
(2) Its diameter.
(3) Its density.
(4) The stretching force.

The laws (expressed with reference to the vibration-number) are as follows:—

I. The vibration-number is inversely proportional to the length of the string.

II. It is inversely proportional to the diameter.

III. It is directly proportional to the square root of the stretching force applied.

IV. It is inversely proportional to the square root of the density of the string.

All the laws are included in the equation

$$n = \frac{1}{2rl}\sqrt{\frac{F}{\pi d}},$$

which gives the vibration-number of a string of radius $r$, and length $l$, and density $d$, stretched by a force $F$.

The violin supplies us with a general illustration of all the above laws. The first law is illustrated by the way in which the violinist 'fingers' a string—shortening or lengthening the vibrating portion so as to get different notes out of the same string. The second law is illustrated by the different thicknesses of the three upper strings. The third law is illustrated by the way in which the violin is tuned, viz. by twisting each string about a tightening-peg. The fourth law is indirectly illustrated by the construction of the lowest or bass string, which is wrapped round with metal wire.

52. The **Sonometer**.—For the purpose of verifying the above laws we make use of the sonometer—an instrument which is so called because it can be used for measuring the pitch of a sound. Fig. 16 shows a sonometer consisting of a

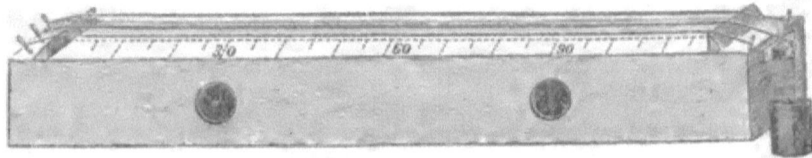

Fig. 16.—Sonometer.

hollow sounding-box on which are stretched one or more strings (or wires). Under the ends of the strings are placed wedge-shaped pieces of hard wood called 'bridges': one of these should be movable, so that the length of the vibrating portion of the string can be altered at will. For the purpose of verifying Law III, one string should be stretched by weights hanging over a pulley. It is convenient to make the full length between the bridges just a metre, and to have a metre scale on the sonometer.

A sonometer suitable for the following experiments can be made of a board 3 ft. 6 in. long, 4 in. broad, and ¾ in. thick. This should be firmly fixed to a wall, nearly vertical, but with the bottom sloping slightly outwards, so as to make the wires bear against the lower bridge. The upper (fixed) bridge is made by bevelling a piece of hard wood 3 in. long and 1 in. square, and facing the sharp edge with brass wire. Two smaller movable bridges of the same kind are made for the lower ends. Two fine piano-wires or violin-strings are fixed to the board above the upper bridge: the lower end of one of these is attached to an iron screw or 'wrest-pin,' which is used for tightening the wire. This wire can be used as a standard of reference, *e.g.* by tuning it to unison with a fork. To the lower end of the other wire is attached a hook or strong scale-pan, on which weights are placed: or a bucket into which water is poured may be used instead of weights.

**53. Verification of the Laws.**—We can now proceed to verify by experiment the first and third laws stated in Art. 51. The law of lengths can be tested with both ease and accuracy.

EXPT. 20. By adjusting the weights on the one wire or the tension on the other, tune the two to unison, using the full length of the wire (100 cm.) Call the note d. Consider how much the wire ought to be shortened to give the note m (a major third above d). The interval between the two notes, or ratio of their vibration-numbers, is $\frac{5}{4}$. Now the law says that the pitch is inversely proportional to the length of the wire. Hence it ought to be shortened to four-fifths of its original length to give the note m, and $\frac{4}{5}$ of 100 is 80. Shift the movable bridge until the length of the vibrating portion of the wire is 80 cm. Pluck the wire and compare the sound with that given by the reference wire (d). You will find that the note is m.

Try the other notes of the major chord (s and d'). You will find that s is given when the length is 66⅔ cm. (100 × ⅔ = 66⅔), and d' when the length is just 50 cm. (100 × ½ = 50).

It will be a good exercise for you now to tune the wire to any note in the scale by ear only, comparing it from time to time with the reference note (or tonic), and shifting the bridge up or down till you exactly hit the note you want: *then* measure the length and compare it with that required by theory. Go through all the notes of the scale in this way and enter up your results in your note-book in three columns, the first giving the name of the note, the second the length of wire found by trial, and the third the calculated length.

You will easily see from the values of the intervals given in Art. 46 that the lengths required for the various notes (the fundamental being taken as 100 cm.) should be

| d | r | m | f | s | l | t | d¹ |
|---|---|---|---|---|---|---|---|
| C | D | E | F | G | A | B | c |
| 100 | 88·8 | 80 | 75 | 66·6 | 60 | 53·3 | 50. |

It is only as a matter of convenience that we have taken the length of the string to be 100 cm. to start with. For any given interval the lengths would be found to have a constant ratio, whatever the original length of the string.

EXPT. 21.—Shift the bridge step by step so as to reduce the length of the wire to $\frac{1}{2}$, $\frac{1}{3}$, $\frac{1}{4}$ . . . . Observe that the notes produced form the harmonic series described in Art. 49.

EXPT. 22.—The following experiment may now be made to illustrate the relation between pitch and stretching force (the length being kept constant).

First hang a 7-lb. weight to the wire, and then substitute for this a 28-lb. weight. The note produced in the second case is an octave above the first. This is clearly in accordance with the law, for the weights are in the ratio of 1 to 4: and $\sqrt{1} : \sqrt{4} = 1 : 2$.

Again, hang on the wire in succession weights of 4 lbs. and 9 lbs. (or in this proportion). The square roots of these numbers are 2 and 3. The interval $\frac{3}{2}$ is a fifth (d : s). Try whether the notes bear this relation to one another.

54. **Other Modes of Transverse Vibration: Overtones.**—We have as yet considered only one mode of vibration of a string, viz. that in which it vibrates as a whole, as indicated in the accompanying figure (Fig. 17, I). It then produces its lowest or fundamental note. But any stretched string may be made to divide up into a number of vibrating segments. For example, suppose the string to be lightly touched in the centre, so as to hinder or 'damp' the vibration at this point. If now it is

bowed at a point midway between the centre and one end, it will divide into two vibrating (or ventral) segments (Fig. 17, II)

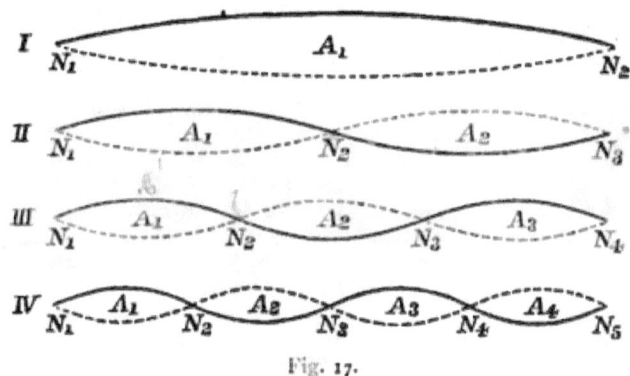

Fig. 17.

separated by a *node*. In Fig. 17, II there are three nodes, $N_1$ and $N_3$ at the ends, and $N_2$ at the centre. Midway between the nodes (at $A_1$ and $A_2$) come the points at which the amplitude of vibration is greatest: these are the *antinodes*.

Again, if the string be damped at one-third of its length, and bowed midway between this and the nearer end, it will divide up into three vibrating segments, as shown in Fig. 17, III.

The existence of nodes and antinodes can be shown, and their positions found, by hanging over the string light paper riders (Fig. 18). These are jerked off at the antinodes, but remain undisturbed (or nearly so) at the nodes.

Fig. 18.

When a string is made to vibrate successively in 1, 2, 3, 4 . . . vibrating segments, the notes produced form the harmonic series referred to in Art. 49. For when the string divides up into 2, 3, 4 . . . segments, each may be regarded as being virtually a separate string of the same material, and stretched by the same force, but of $\frac{1}{2}$, $\frac{1}{3}$, $\frac{1}{4}$ . . . the length of the whole string. Hence by our first law the vibration-number is 2, 3, 4 . . . times that of the note produced when the string vibrates as a whole. The latter is called the fundamental and the others the *overtones*.

EXPT. 23.—Tune the two strings on the sonometer to unison again. The length is not of importance, but we shall assume it to be 100 cm. Call the note produced d. Damp one of the strings by touching it lightly with a feather at the centre (50). Bow it midway between this and the end. It now gives the note $d^1$, an octave above the first.

Damp it at $33\frac{1}{3}$ cm., corresponding to the point $N_2$ in Fig. 17, III. Bow it midway between this and the end (at $A_1$). It gives the note $a^1$.

Repeat this, having first placed light paper riders at $A_2$, $N_3$, and $A_3$. After a little practice you will easily unhorse the first and last without disturbing the second.

By damping the string at 25 cm., 20 cm., and $16\frac{2}{3}$ cm., the overtones $d''$, $m''$, and $s''$ can be obtained.

**55. Quality of Tone.**—Overtones are almost invariably present together with the fundamental note produced by a string, *and upon the number and relative strength of the overtones depends the quality of the tone produced.*

Thin and tightly stretched metallic wires easily give a large number of overtones, and the presence of these can easily be detected by the unaided ear, even when the string is not specially damped at any point. High overtones are more readily produced by bowing a string than by striking it with a soft hammer or plucking it with the finger: more readily still by plucking with the finger-nail or striking with a hard-pointed instrument, when the quality of the sound becomes poor and 'tinkling.' This effect can be observed in pianos as the soft padding of the hammers wears out.

Again, the point at which a string is plucked or bowed exercises an important effect on the overtones produced; for any overtone which has a node at the point must necessarily be absent. In pianos the hammers are generally made to strike each string at one-seventh the length from one end: the note produced includes all harmonics up to the sixth.

### Examples on Chapters V-VI.

1. What variety of notes can you get out of a stretched string, without altering its tension or length? What will be the effect of halving its length by a fixed bridge?

2. Describe an instrument by which the pitch of the notes emitted by two vibrating strings may be compared. If they were different, how would you attempt to bring them into unison?

3. State in what way the rate of transverse vibration of a stretched string depends upon the tension. How would you determine the rate of vibration of the string?

4. Explain the character of the vibration of a stretched violin string. What effect is produced by touching it at one-third of its length from one end?

5. A and B are two wires of the same material and thickness. A is 2 feet long, and is stretched by a weight of $8\frac{1}{2}$ pounds. B is 4 feet long, and stretched by a weight of 34 lbs. How are the notes which the wires yield when struck related to one another?

6. A steel wire, one yard long, and stretched by a weight of 5 lbs., vibrates 100 times per second when plucked. If I wish to make two yards of the same wire vibrate *twice* as fast, with what weight must I stretch it?

7. A vibrating string is found to give the note $f$ when stretched by a weight of 16 lbs. What weight must be used to give the note $a$? and what additional weight will give $c^1$?

# CHAPTER VII

## RESONANCE

**56. Forced and Free Vibrations.**—We have already seen that the sound produced by a tuning-fork is greatly strengthened by the use of a sounding-box. The same principle is employed in the construction of violins and pianos. The sounding-boards of such instruments are made of thin elastic wood, which admits of being thrown into a state of *forced vibration* corresponding to any note within the range of the instrument.

But, in general, every elastic body has certain modes of *free vibration* which are natural to it, and can be easily excited in it by even slight impulses *if they are repeated at the proper intervals.*

We may illustrate the difference between the two things as follows. A child who knows how to manage a swing can set it swinging through a large arc without any great effort by timing his impulses so as to coincide with the natural period of vibration of the swing. On the other hand, if he wriggles about irregularly, one impulse may destroy the effect of the preceding one, and the result is simply a slight and fitful jerking.

This principle of the accumulation of small impulses enables us to explain a number of phenomena which are classed under the general heading of *resonance* or *co-vibration.*

**57. Sympathetic Vibration of Tuning-Forks.**—One of the most striking instances of co-vibration is afforded by the fact that a vibrating tuning-fork is able to throw into vibration another fork in its neighbourhood, *provided both are adjusted to exactly the same pitch.* The two forks should be placed with the open ends of their sound-boxes facing each other (Fig. 19). One of them should be strongly bowed and then brought to rest by touching it with the finger. On placing the ear close to the other fork the same note will be heard. But for class purposes it is best to place in contact with one prong of the second fork a light indicator, such as a microscope cover-

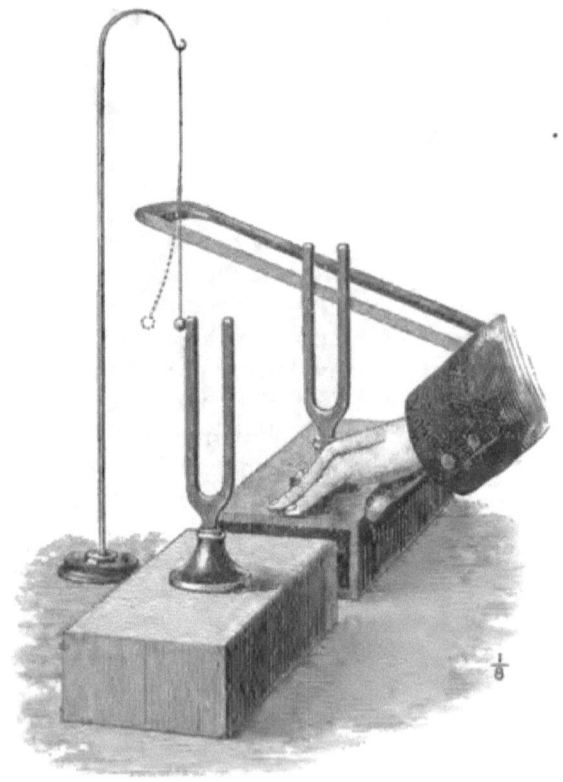

Fig. 19.

glass or small cork ball (varnished to make its outer surface hard), suspended by a silk fibre.

If one of the forks be made to vibrate more slowly by loading its prongs with wax, the experiment no longer succeeds. This shows that it is essential to have the two forks exactly in unison. The energy given out by the first fork is carried through the air in the form of sound-waves, and taken up by the second fork.

**58. Resonance of Air-Columns.**—An easier way of illustrating the phenomena of resonance is by making a tuning-fork set a column of air into vibration.

EXPT. 24.—All that is required is a tuning-fork, some water, and a narrow cylindrical vessel such as a gas-jar (Fig. 20), or a glass tube that can be lowered into water. If the fork is a C fork (256 vibrations per second), the gas-jar should not be less than about 40 cm. deep.

Strike the fork and hold it over the mouth of the jar; there will be some increase in the loudness of the sound. Gradually pour water into the jar; when it reaches a certain depth the sound swells out in a marked manner. When this point is overstepped the resonance gradually diminishes.

Try another fork of different pitch: you will find that the length of the air-column which gives the best resonance is greater or less according as the pitch is lower or higher.

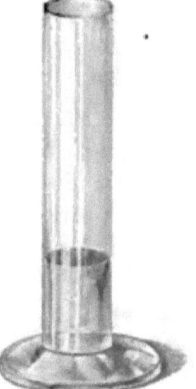

Fig. 20.

Thus there is a certain length of air-column which most loudly reinforces the note of a given fork: and conversely an air-column of given length is most easily thrown into vibration by a note of given pitch, and can be used as a 'resonator' for reinforcing that note or detecting its presence.

# CHAPTER VIII

## VIBRATION OF AIR-COLUMNS—ORGAN-PIPES, ETC.

**59. Stationary Vibration of Spiral Spring.**—We have already made use of the wave-machine described on p. 238 to illustrate how *progressive* waves (or successive pulses of condensation and rarefaction) travel along air in a cylindrical tube, and how they are reflected (p. 250). We can also use the same spiral spring to illustrate how successive pulses of compression and rarefaction applied at proper intervals to one end of a tube are able to set the air within it into a state of *stationary vibration*.

EXPT. 25.—Fix one end of the spiral, say the left-hand end, as in Expt. 11 (p. 250). Start a pulse of compression from the right-hand or free end. It is reflected back, still as a compression, from the fixed end. When this pulse reaches the free end the outer coils, having no pressure from other coils in front to resist their motion, swing freely outwards (to the right). Meanwhile the coils just behind them are beginning to swing back (to the left), so that at the free end the pulse of compression is converted into a pulse of rarefaction, or reflected *with change of type*. The pulse of rarefaction travels along the coil, is reflected without change of type (but with change in the direction of motion) at the fixed end, and in turn is reflected with change of type from the free end. This process goes on until the motion gradually dies out.

Next, suppose that instead of a single pulse we have a series of alternate pulses of condensation and rarefaction sent along from the free end. Corresponding to this series of incident waves there will be a series of reflected waves travelling in the opposite direction from the fixed end. The displacement at any point of the spiral will be the resultant of the displacements due to each set of waves separately. If both tend to produce displacements in the same direction, the resultant will be the sum of the separate displacements: if both tend to produce displacements in opposite directions, the resultant will be the difference of the separate displacements. The two sets of waves are said to *interfere* with one another. In general the interference is partial. When the separate displacements are equal but in opposite directions, the interference is complete, and the particle or coil remains at rest.

The behaviour of our spiral coil under the influence of the two sets of waves will clearly depend upon where and under what circumstances they meet. Under certain conditions the coil may be put into a state of stationary (longitudinal) vibration corresponding to the stationary (transverse) vibration of a stretched string. When this happens it is clear that the fixed end must be a node, for no motion is possible there; and we may reasonably expect the antinode to be at the free end.

The necessary conditions may be found as follows. Take hold of the right-hand (free) end, and send a gentle pulse of condensation along the coil. Without removing your hand, watch the reflected pulse of condensation, and, just at the instant when it arrives at the free end, pull the coil outwards as in producing a pulse of rarefaction. Then again push it inwards just in time to meet the reflected pulse of rarefaction, and so on. After a little practice you will be able to throw the coil into a steady state of stationary vibration, which lasts for some time after your hand is removed. Carefully examine the nature of the motion. Notice that at the fixed end the coils are alternately crowded together and then drawn apart: these changes of relative position correspond to changes of density in the case of air. At the free end there is no change of density, but the amplitude of vibration is greater than at any other part of the spiral: the coils swing freely backward and forward, but remain at the same distance apart. You have produced a stationary vibration with a node at the fixed end and an antinode at the free end.

*A node is a place* ***where there is no motion,*** *but where the changes of density are greatest.*

*An antinode is a place where there is no change of density, but where the amplitude of vibration is greatest.*

**60. Both Ends Free.**—The spiral coil can also be thrown into a state of stationary vibration when both its ends are free. The ends now are antinodes and the centre is a node: for, as will be shown by the following experiment, the two sets of waves travelling in opposite directions always produce complete interference at the centre, so that it remains permanently at rest.

EXPT. 26.—Take hold of both ends of the spiral and push them gently inwards. You thus produce two pulses of condensation which meet in the middle and cross over to either end. Just when these arrive at the ends, assist the motion by pulling both ends outwards. After repeating this process two or three times you can take your hands away from the coil, and its two halves will go on steadily vibrating in and out. The motion may at first sight remind you of the way in which a concertina opens and shuts: but there is this essential difference. The rarefactions (when the ends swing outwards) and condensations (when they swing inwards) are not uniformly distributed along the spiral. They are greatest at the centre, which remains stationary; whereas at the ends, where the motion is greatest, the coils always remain at the same distance apart. The vibration takes place in such a way that there is a node at the centre and an antinode at each end.

Observe that the rate of vibration is twice as great as in the last experiment (one end fixed).

**61. Wave-Length.**—In the case of a spiral with one end fixed, the wave-length is *four times* the length of the spiral. During the time of a complete vibration the wave travels four times across the spiral. First (as the outer coils move inwards) there is a pulse of condensation, which travels to the fixed end and is then reflected back to the free end. Next (as the outer coils swing outwards) begins a pulse of rarefaction, which similarly travels to the fixed end and back. Matters are now in the same state as at the start, and the same process is repeated.

When both ends are free the wave-length is *twice* the length of the spiral. For, on account of the change of type which occurs in reflection of a pulse from the free end of a spiral, the pulse has only to travel twice across the spiral (to the end and back) in order to get back to the condition in which it started.

**62. Resonance of Air-Columns.**—The resonance of an air-column (Art. 58) is due to a stationary vibration of the air produced by the tuning-fork or other source of sound. If the column of air is contained in a tube closed at one end, the air moves just in the same way as the spiral in Expt. 25.

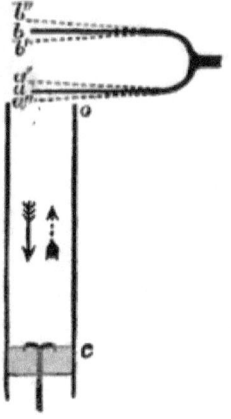

Fig. 21.

Suppose one prong of the tuning-fork to be moving downward from $a'$ to $a''$ (Fig. 21). In so doing it produces a pulse of condensation which runs down the tube and is reflected up from the closed end. From what has been already stated it will be clear that the necessary condition for reinforcement of the sound is that this reflected pulse should reach the open end of the tube just as the prong begins its return journey from $a''$ to $a'$; the air will then be moving in the same direction as the prong. Notice that the reflected pulse of condensation reaches the open end just as the fork is beginning to send down a pulse of rarefaction; owing to the interference of these two the density of the air at the open end remains unchanged. Now a pulse of rarefaction starts down, is reflected from the closed end, and meets the prong just after it has completed a vibration. It is now once more starting from $a'$ to $a''$ and sending down another pulse of condensation. In the time required for a complete vibration we have had a pulse of condensation and then one of rarefaction, each travelling twice the length of the tube. Imagine the bottom of the tube knocked out and you will easily see that in the interval between two successive and similar pulses (say of condensation) the wave would have travelled forward through a distance equal to *four times the length of the tube*. This, then, is the wave-length of the note which the tube reinforces. If $l$ be the length of the tube, the wave-length is $\lambda = 4l$.

In the case of a tube open at both ends the air moves like the spiral in Expt. 26. The wave-length of the note which it reinforces is only *twice the length of the tube*, or $\lambda = 2l$. Hence an open tube which is to resound to a given note or given tuning-fork must be twice as long as the corresponding closed tube. Imagine the closed tube in Fig. 21 replaced by an open tube twice as long, and let the prong of the fork be moving downwards. In the time that it takes to go from $a'$ to $a''$ the pulse of condensation will have travelled the length of the tube. It is reflected from the open end *with change of type*, *i.e.* as a pulse of rarefaction, and in the time that the prong takes to go from $a''$ to $a'$ this pulse will have travelled to the top of the tube. Here it is reflected as a pulse of condensation which coincides with that produced by the fork as it begins its next vibration. Thus the conditions for resonance are fulfilled. The open ends of the tubes are antinodes and there is a node at the middle of the tube.

Hints for Experiments.—As a closed resonator of adjustable length you may use a glass tube closed at one end by a cork (Fig. 21), which can be pushed in or out until the maximum resonance is obtained. For an adjustable open resonator make a paper tube which will just slip over the glass tube, and slide this in or out as may be required. Verify by experiment the statement that the length of the open resonator is double that of the closed one. For a $c$ fork ($n=256$) you will find that the length of the closed resonator is about 33 cm. while that of the open resonator is about 66 cm. Both results indicate that the wave-length in air of the note $c$ is 132 cm. ($4 \times 33 = 2 \times 66 = 132$).

Assuming that the vibration-number of the fork is correct, you can proceed to calculate from your experiment the velocity of sound in air. For we have seen (Art. 16) that the velocity in any medium is given by the equation $v = n\lambda$. Here $n = 256$, $\lambda = 132$ cm., and $\therefore v = 256 \times 132 = 33{,}792$ cm. per second.

The same equation shows that $n = \frac{v}{\lambda}$. As the velocity in any given medium (say air) is constant and independent of the pitch, it follows that the vibration-number of a given note is inversely proportional to its wavelength in that medium. Hence also the vibration-number (or pitch) of the note to which a tube resounds is inversely proportional to the length of the tube.

Now this note is precisely that which the air-column emits on its own account when it is thrown into a state of stationary vibration. Such vibrations are easily produced by blowing across the edge of a tube,—say a glass tube about 1 cm. in diameter. Take a tube 33 cm. long, close the lower end with your thumb and blow across the upper edge: it gives the note $c$. A tube twice as long open at both ends gives the same note. The latter tube (66 cm.) closed at one end gives the note C, an octave below.

**63. Organ-Pipes.**—Fig. 22 illustrates the construction of a common form of organ-pipe with a 'flue' or 'flute' mouthpiece, which is very much like that of an ordinary whistle. The air passes from the wind-chest through the conical tube at the bottom of the pipe, escapes through a narrow horizontal slit and strikes against the sharp bevelled edge opposite, which is called the *lip*. The air escapes in an intermittent manner, with a rushing noise, due to a mixture of vibrations of different frequencies. The air-column selects out of these the particular one which it reinforces, and when this happens the pipe speaks or emits a note.

The pipe shown in the figure is an *open pipe*. It is open to the air at the bottom (below the lip) as well as at the top, and both of these places are antinodes. There is a node in the middle of the tube (see Fig. 23).

A pipe which is closed at one end (the top) is called a *closed or stopped pipe*. There is always a node at the closed end of the pipe, for the air there cannot move. The open (or mouth) end is an antinode, for the density of the air there remains constant and equal to that of the air outside. The note emitted by a stopped pipe is always an octave below that emitted by an open pipe of the same length.

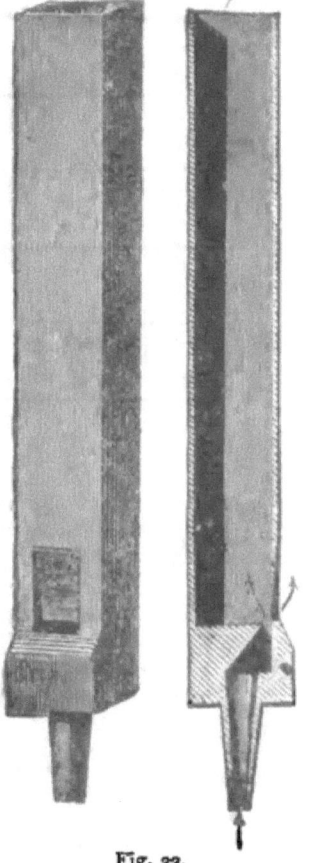

Fig. 22.

**64. Overtones of Pipes—Quality.**—The modes of vibration above described are those of pipes producing their lowest or fundamental note. But by blowing into a pipe more strongly it may be made to 'jump' or produce one or more overtones. The series of overtones produced by overblowing a stopped pipe is not the same as for an open pipe. In discussing the possible overtones in either case we must remember that, as in the case of a string vibrating in segments, the nodes and antinodes follow each other at regular intervals, and that the distance from a node to the nearest antinode is quarter of a wave-length.

*Open Pipes*—(Fig. 23). The ends of an open pipe are always antinodes. When the fundamental note (I) is sounded there is a node

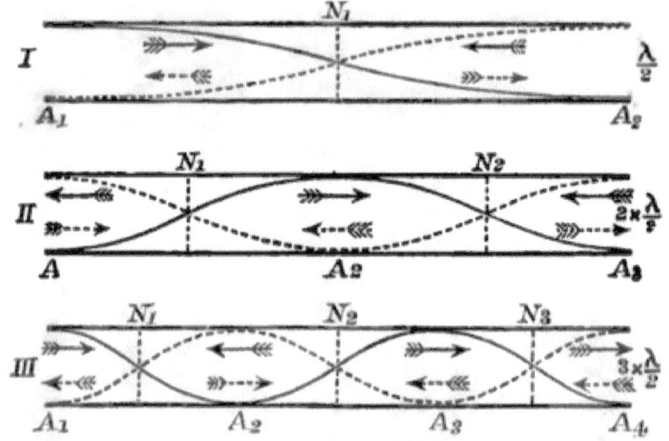

Fig. 23.—Overtones of Open Pipe.

($N_1$) at the centre, and **the length of the pipe is half the wave-length of the note**. When the first **overtone** (II) is produced, a fresh antinode appears at $A_2$ with nodes at $N_1$ and $N_2$. The air-column divides into two equal

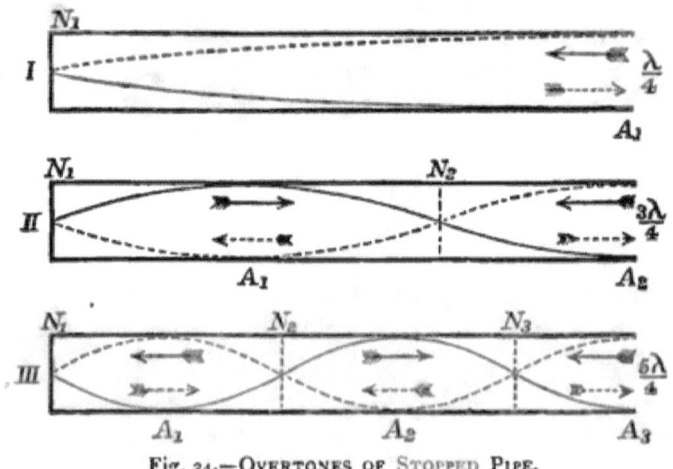

Fig. 24.—Overtones of Stopped Pipe.

vibrating **segments**, and the wave-length is half that of the fundamental. When **the second overtone** (III) **is** produced the column divides into three equal vibrating segments, and the wave-length is one-third that of the fundamental. Thus the possible overtones (together with the fundamental) include the whole harmonic series of Art. 49 (1, 2, 3, 4, 5, 6 . . .).

*Stopped Pipes*—(Fig. 24). In the case of a stopped pipe there must always be a node at the stopped end, and an antinode at the open end. Thus when the fundamental (I) is produced, the length of the pipe is a quarter-wave-length. The even harmonics in the series of Art. 49 (*i.e.* the successive octaves, 2, 4, 6 . . .) are absent from the possible overtones. The first possible division into segments is that shown in Fig. 24, II, with nodes at $N_1$ and $N_2$ and antinodes at $A_1$ and $A_2$; the pipe then includes three quarter-wave-lengths. The next division is shown in Fig. 24, III, when the pipe includes five quarter-wave-lengths. The corresponding vibration-numbers are 1, 3, 5, 7. . . .

Some of the overtones are generally present, together with the fundamental note of any pipe, and upon their number and relative strength depends the quality of the tone produced. We may clearly expect the quality of tone of a stopped pipe to be different from that of an open pipe. Again, the quality depends upon the form of the pipe: *e.g.* a narrow pipe more readily yields harmonics than a wide pipe, especially a wide stopped pipe.

**65. Vibrations of Rods.**—Rods of glass, wood, brass, etc., may be made to vibrate longitudinally and produce musical notes by rubbing them in the direction of their length. A glass rod may be thrown into a state of vibration by drawing a wetted cloth quickly along it; a rod of wood or brass by means of a cloth dusted over with powdered resin. The vibrations produced are stationary vibrations similar to those of organ-pipes and following the same laws. If the rod is clamped at the centre it vibrates like the air in an open pipe with a node at the centre and antinodes at the ends. If it is clamped at one end, there is a node at that end and an antinode at the free end, as in a stopped pipe.

### Examples on Chapters VII-VIII.

1. Upon what physical properties do (1) the loudness, (2) the pitch of a musical note, depend? Two organ-pipes of the same length are one of them open and the other closed; how are their notes related as regards pitch?

2. Describe the state of disturbance of the air in a pipe closed at one end, when it resounds to a tuning-fork which is held over it. State the relation between the length of the pipe, the pitch of the note, and the velocity of sound in air.

3. State how the air moves in different parts of a tube 1 ft. long, open at both ends, when sounding its fundamental note. What note does it give?

4. A glass rod 5 ft. long is clamped at its centre, and rubbed longitudinally with a wet cloth. State how it vibrates when thus treated, and calculate the velocity of sound in the glass, if told that the above rod makes 1295 complete vibrations every second.

5. Find the length of a closed organ-pipe which when blown at 15° gives the note *c* (256 vibrations per sec.)

# ANSWERS TO EXAMPLES

### CHAPTERS I-IV (p. 254)

**1.** See pp. 24, 25, 233. **2.** See Art. 10. **3.** See Arts. 23, 37. **6.** See pp. 246, 249.

### CHAPTERS V-VI (p. 268)

**5.** They yield the same note. **6.** 80 lbs. **7.** 25 lbs.; an additional 11 lbs. (*i.e.* total weight = 36 lbs.)

### CHAPTERS VII-VIII (p. 279)

**1.** See Arts. 7, 10, 40, 63. **2.** See Arts. 62, 63. **3.** The note depends upon the velocity of sound in air, and this again upon the temperature. If we take it to be 1116 ft. per second (p. 246) the vibration-number of the note would be 558. **4.** The wave-length of the note in glass is 10 ft. (twice the length of the rod). The velocity in glass is $v = n\lambda = 1295 \times 10 = 12,950$ ft. per second. **5.** 1·09 ft. or 33·2 cm. See Ex. 3 and p. 246.

# INDEX

[*Numbers refer to Pages*]

Absolute temperature, **32, 33**
Absorption and radiation, **103, 116**
Absorptive power, **101**
Air-thermometer, **34**
Air-thermoscope, 5-7
Amplitude of vibration, **230**
Angle, critical, 179
Antinodes, 267, 274, 278
Aqueous vapour in atmosphere, **72**
Atmospheric pressure, **28**

Barometer, 28
Boiling, laws of, 68
Boiling-point, 56, 60
Boyle's law, 29
Breezes, land and sea, 106
Bunsen's photometer, 127

Camera, photographic, 213
Capacity, thermal, **45**
Centigrade scale, **10**
Charles's law, 31
Chord, common **or** major, **258**
Clouds, 108
Compensation of clocks and **watches, 18**
Conduction, 79
Conjugate foci: of mirror, **151**; of lens, 196
Convection, 79, 89
Critical angle, 179
Cryophorus, 69

Daniell's hygrometer, 75
Davy lamp, 84
Density, 24, 25
Dew, **108**; dew-point, 75
Diathermancy, 104
Diatonic scale, 257
Dines's hygrometer, **76**
Dispersion, 218

Echoes, **251**
Eclipses, **121**
Elasticity, 233
Emissive power, 103
Evaporation, 56; cooling effect of, **68-70**
Expansion, 4; linear, 14; coefficient **of,** 16; superficial and cubical, 20; **real** and apparent, 22; of water, 24; **of gases, 27-36**; on freezing, 53
Eye, **213**

Fizeau (velocity of light), **132**
Flames, sensitive, 251
Focal length: of mirror, **149,** 161; of lens, 194, 205
Freezing mixtures, 52
Fusion, 48; latent heat of, 50

Gases, properties of, 27
Gulf Stream, 96

Harmonics **or** overtones, 261, 266, **268**
Heat, specific, 40; unit of, **40**
Hope's experiment, 25
Humidity, 74
Hygrometers, Daniell's, 75; **Dines's, 76**; wet and dry bulb, 77

Ice-Calorimeter, **51**
Images produced **by** small apertures, 118; by plane mirrors, 142; by spherical mirrors, 155-166; by lenses, 199-**210**
**Index of** Refraction, 172
Intensity of illumination, 123; **of sound,** 235, 244
Intervals, musical, 257-260
Inverse squares, law of, **124**

Lantern, optical, **212**

# INDEX

Laplace's correction, 248
Latent heat, 49; of water, **50**; **of steam,** 65
Laws: Boyle's, **29**; Charles's, 31; **of** ebullition, 68; of inverse squares, 124; of reflection, **137**; of refraction, **172**
Lenses, **191**; convex, 199; concave, **209**
Light, 115; rectilinear propagation **of,** 117; velocity of, 131; reflection **of,** 135; refraction of, 169

MAXIMUM pressure of vapour, 58-60
Melting-point, 48
Microscope, simple, 207; compound, 215
Minimum deviation, 189
Mirrors, plane, 136; spherical, 148; concave, 149; convex, 165
Musical intervals and scale, 257

**NEWTON'S** experiment, 217; formula, 248
**Nodes, 267,** 273

Opera-glass, 216
Optical bench, **129**; **centre, 193**; lantern, 212
Organ-pipes, 276
Overtones or harmonics, 261, **266**; of strings, 267; of pipes, 278

PENDULUM, compensated, 18
Phase, 240
Photometry, 123
Pitch, 234, 255
Principal focus: of mirror, 149; **of lens,** 194

QUALITY of tone, 268, 279

RADIATING power, 103

Radiation, **79, 97, 115**
Rain, 109
Reflection of radiant heat, 99; of light, **135**; total, 181; of sound, 251
Resonance, 269; of air-columns, **271, 275**
Römer (velocity of light), 131
Rumford's photometer, 125

SAFETY-LAMP, **84**
Saturated vapour, **59**
Scale, diatonic, **257**
Sensitive flames, 251
Shadow, 120: photometer, 125
Siren, 256, 259
Sonometer, 264
Specific heat, 40
Spectrum, 217; pure, 219
Stationary vibrations, 263, 272
Steam, latent heat of, 65
Strings, vibrations of, 263

TELESCOPE, 214
Temperature, 3, 39; absolute, 33
Thermal capacity, 45
Thermometer, 8: differential, 34
Total reflection, 181
Trade winds, 107
Transmissive power, 104

**UNIT of heat, 40**

VAPOUR-PRESSURE, **57**; maximum, 58
Velocity of light, 131; of sound, 245
Vibration, 230; stationary, 263, 272
Virtual images produced by mirrors, 141, **159,** 166; by lenses, 207, 209

WATER, maximum density of, 24-26; specific heat, 46; latent heat, 50
Wave-line, 234; wave-length, 239, 276
Weight thermometer, 23

THE END

*Printed by* **R. & R.** CLARK, *Edinburgh.*

www.ingramcontent.com/pod-product-compliance
Lightning Source LLC
Chambersburg PA
CBHW032055230426
43672CB00009B/1598